The Cosmos Symmetries – Gravitation Connection

GravityElectroWeakStrong (GEWS) Theory
GEWS Field Equations from ElectroWeakStrong
Finite Gravitation Perturbation Theory
Time-Space Theory
The PseudoTime of the Cosmos
Black Hole Event Horizon Time-Space

Stephen Blaha Ph. D.
Blaha Research

MMXXVI

ISBN: 979-8-9950914-5-5

August 10, 2026

To Margaret

Some Other Books by Stephen Blaha

SuperCivilizations: Civilizations as Superorganisms (McMann-Fisher Publishing, Auburn, NH, 2010)

Unification of God Theory and Unified SuperStandard Model THIRD EDITION (Pingree Hill Publishing, Auburn, NH, 2018).

The Exact QED Calculation of the Fine Structure Constant Implies ALL 4D Universes have the Same Physics/Life Prospects (Pingree Hill Publishing, Auburn, NH, 2019).

Passing Through Nature to Eternity ProtoCosmos, HyperCosmos, Unified SuperStandard Theory (Pingree Hill Publishing, Auburn, NH, 2022).

HyperCosmos Fractionation and Fundamental Reference Frame Based Unification: Particle Inner Space Basis of Parton and Dual Resonance Models (Pingree Hill Publishing, Auburn, NH, 2022).

The Cosmic Panorama: ProtoCosmos, HyperCosmos, Unified SuperStandard Theory (UST) Derivation (Pingree Hill Publishing, Auburn, NH, 2022).

God and and Cosmos Theory (Pingree Hill Publishing, Auburn, NH, 2023).

Newton's Apple is Now The Fermion (Pingree Hill Publishing, Auburn, NH, 2023).

Cosmos Theory: The Sub-Particle Gambol Model (Pingree Hill Publishing, Auburn, NH, 2023).

Cosmos-Universe-Particle-Gambol Theory (Pingree Hill Publishing, Auburn, NH, 2024).

Structure and Dynamics of Cosmos Theory and the Unified SuperStandard Theory (Pingree Hill Publishing, Auburn, NH, 2024).

Black Holes, White Holes, and Superluminal Starship (Pingree Hill Publishing, Auburn, NH, 2025).

The Sequence Theory Foundation of Cosmos Theory and Gambol Theory (Pingree Hill Publishing, Auburn, NH, 2025).

Fractal Cosmos (Blaha Research, Auburn, NH, 2025)

Fractal Cosmos II (Blaha Research, Auburn, NH, 2025)

Fractal Cosmos III (Blaha Research, Auburn, NH, 2025)

Fractal Cosmos IV (Blaha Research, Auburn, NH, 2025)

The Entropic Basis of Cosmos Theory The Agent of Origin (Blaha Research, Auburn, NH, 2025)

Cosmos Proof of the Existence of God, (Blaha Research, Auburn, NH, 2026)

Available on bn.com, Amazon.com, Amazon.co.uk and other international web sites as well as at better bookstores.

CONTENTS

Introduction

Einstein developed General Relativity (GR) based on Newtonian Gravity and on considerations of simplicity. This book develops a new General Relativity called Enhanced Extended General Relativity (EEGR) based on the natural agreement of the structure of the author's ElectroWeakStrong (EWS) Theory and General Relativistic features. The agreements are major: Riemann Curvature and the Strong Interactions; a new vector field, B, and ElectroWeak Theory; two Higgs fields and two gravitational Higgs vectors; and so on.

The basis in EWS theory recommends itself since ElectroWeak and Strong Interaction features are well known while the features of the "simple" field equations of Gravitation are problematic.

The result was EEGR, which has marvelous new features compared to conventional GR: an explanation of the [mass^2] dimension of the gravitation constant G, the increase of time in the universe through a Time-Space Theory, finding the long sought Absolute Reference Frame, finite Quantum Gravity Perturbation Theory, and a Black Hole event horizon with a constant time (not increasing) and space.

The book describes the author's Time-Space Theory (an offshoot of EEGR) that explains one of the great mysteries: Why does time increase – always and everywhere? The theory considers increasing time and space (Why doesn't all space distances increase like time?), time loops, time spurts, time-space origin of Cosmic strings, frozen time, jumps in time, variations in time – faster and slower, explosive growth in time, collapse in time, and possible Astrophysical evidence for Time-Space Theory.

The book also applies the author's PseudoQuantum Field Theory (developed in the 1970s, and Two-Tier Theory (developed in the first decade of this century) to EEGR and its Perturbation Theory. EEGR perturbation theory has no infinities.

A major event for Gravitation and Cosmos Theory!

1. "Mirroring" The New Gravitation and Cosmos Theory Symmetry Groups

Recently[1] we developed a new Extended General Relativity (EGR) that accounted for many known properties of Gravitation: the [mass^2] dimension of G, the increase of time in the universe, the Absolute Reference Frame, and so on.

We had also previously built a Lagrangian for the Unified SuperStandard Theory (UST) which describes the particles and symmetries of our universe. The UST was described in part by the ElectroWeakStrong (EWS) Lagrangian.

We found that there was a strong similarity – a mirroring – between the EWS parts of the UST: SU(2)⊗U(1)⊗SU(4) and the EGR. We describe this similarity, which has the form of a mapping, in detail in chapter 4. A strong resemblance was found between their structures. This resemblance led us to the Enhanced Extended EGR – the EEGR – described in detail in chapter 4.

We see that the mirroring is ultimately based on the derivation of the origin of Cosmos Theory – in particular, the relation between dimension arrays and the form of higher dimension fermion quantum fields.

The relationship stems from a mapping of Cosmos Theory symmetry group indices to space-time coordinates. We begin with the ElectroWeakStrong (EWS) Lagrangian. Then describe an enhanced Lagrangian version of our EGR presented in Blaha (2026d). These Lagrangians will be seen to be related by the mapping.

Subsequently, we describe additional topics including Universe entropy, time increase, Time-Space Theory, Cosmos spaces, entropy, and other topics such as finite EEGR perturbation theory.

---

[1] Blaha (2026d).

2. EWS Lagrangian For Cosmos Symmetry Groups

2.1 ElectroWeakStrong (EWS) Lagrangian[2]

In this chapter we present the EWS part of the PseudoQuantum formulation of the UST Lagrangian that we previously developed. It uses its ElectroWeak and Strong interaction terms, and the relevant boson Higgs fields terms, to specify the unified Higgs-Confinement Mechanism. This $\mathscr{L}_{EWS}$ Lagrangian generalizes directly to all four generations and four layers of the UST.[3]

$$\mathscr{L}_{EWS} = (\partial_\mu \varphi_1{}^*)(\partial^\mu \varphi_2) + (\partial_\mu \varphi_{1s}{}^*)(\partial^\mu \varphi_{2s}) - ½W_1{}^{\mu\nu}W_{2\mu\nu} - ½F_1{}^{a\mu\nu}F_2{}^a{}_{\mu\nu} - \qquad (2.1)$$
$$- (W_2{}^\mu\varphi_1)^\dagger W_{2\mu}\varphi_1 - (W_1{}^\mu\varphi_2)^\dagger W_{1\mu}\varphi_2 - (A_1{}^{a\mu}\varphi_{1s})^\dagger A_1{}^a{}_\mu\varphi_{1s} - (A_2{}^{a\mu}\varphi_{2s})^\dagger A_2{}^a{}_\mu\varphi_{2s} -$$
$$- V(\varphi_1) - m_1{}^2\varphi_1{}^*\varphi_2 - ½m_2{}^2\varphi_1{}^*\varphi_1 - ½\, m_3{}^2\varphi_2{}^*\varphi_2 -$$
$$- V(\varphi_{1s}) - m_{1s}{}^2\varphi_{1s}{}^*\varphi_{2s} - ½m_{2s}{}^2\varphi_{1s}{}^*\varphi_{1s} - ½\, m_{3s}{}^2\varphi_{2s}{}^*\varphi_{2s} + CC$$

where there are two Higgs fields φ_i with i = 1, 2 for the ElectroWeak sector and two Higgs fields φ_{is} with i = 1, 2 for the Strong sector. The gauge fields $W_1{}^{\mu\nu}$ and $W_{2\mu\nu}$ are the SU(2)⊗U(1) ElectroWeak fields. The Higgsian mass terms are $m_{1s}{}^2\varphi_{1s}{}^*\varphi_{2s} - ½m_{2s}{}^2\varphi_{1s}{}^*\varphi_{1s} - ½\, m_{3s}{}^2\varphi_{2s}{}^*\varphi_{2s}$ + and $m_{1s}{}^2\varphi_{1s}{}^*\varphi_{2s} - ½m_{2s}{}^2\varphi_{1s}{}^*\varphi_{1s} - ½\, m_{3s}{}^2\varphi_{2s}{}^*\varphi_{2s}$ for the respective sectors.

We now introduce explicit symmetry group indices for use later.

$$\mathscr{L}_{EWS} = (\partial_\mu \varphi_{1b}{}^*)(\partial^\mu \varphi_{2b}) + (\partial_\mu \varphi_{1sa}{}^*)(\partial^\mu \varphi_{2sa}) - ½W_{1b}{}^{\mu\nu}W_{2b\mu\nu} - ½F_{1a}{}^{\mu\nu}F_{2a\mu\nu} - \qquad (2.2)$$
$$- (W_2{}^\mu\varphi_1)_b{}^\dagger(W_{2\mu}\varphi_1)_b - (W_1{}^\mu\varphi_2)_b{}^\dagger(W_{1\mu}\varphi_2)_b - (A_1{}^\mu\varphi_{1s})_a{}^\dagger(A_1{}^a{}_\mu\varphi_{1s})_a - (A_2{}^\mu\varphi_{2s})_a{}^\dagger(A_{2\mu}\varphi_{2s})_a -$$
$$- V(\varphi_1) - m_1{}^2\varphi_{1b}{}^*\varphi_{2b} - ½m_2{}^2\varphi_{1b}{}^*\varphi_{1b} - ½\, m_3{}^2\varphi_{2b}{}^*\varphi_{2b} -$$
$$- V(\varphi_{1s}) - m_{1s}{}^2\varphi_{1sa}{}^*\varphi_{2sa} - ½m_{2s}{}^2\varphi_{1sa}{}^*\varphi_{1sa} - ½\, m_{3s}{}^2\varphi_{2sa}{}^*\varphi_{2sa} + CC$$

where the indices labeled "a" run from 1 through 16 and where the indices labeled "b" run from 1 through 4 and where CC represents Complex Conjugate. There are four $W_{ib}{}^\mu$ fields and sixteen $A_{1a}{}^\mu$ fields.

The condensate potential that leads to Strong SU(4) gluons supporting quark confinement is:

$$V(\varphi_{1s}) = \sum_{a=1}^{16} \lambda_{\varphi s}(\varphi_{1sa}{}^*\varphi_{1sa} - \varphi_{1s0a}{}^2)^2$$

where

$$\varphi_{1s00} = 0$$

[2] From Blaha (2026b) and (2026c).

3 We omit certain additional terms that normally appear in PseudoQuantum Lagrangians: $¼\,(\partial_\mu\varphi_{1b}{}^*)(\partial^\mu\varphi_{1b})$ + $+ ¼(\partial_\mu\varphi_{1sa}{}^*)(\partial^\mu\varphi_{1sa}) - ¼W_{1b}{}^{\mu\nu}W_{1b\mu\nu} - ¼F_{1a}{}^{\mu\nu}F_{1a\mu\nu}$.

$$\varphi_{1s0a} = \varphi_{1s0} \qquad \text{for } a = 2, 3, \ldots, 16$$

and

$$V(\varphi_1) = \sum_{b=1}^{4} \lambda_\varphi(\varphi_{1b}{}^*\varphi_{1b} - \varphi_{10b}{}^2)^2$$

where

$$\varphi_{100} = 0$$
$$\varphi_{10b} = \varphi_{10} \qquad \text{for } b = 2, 3, 4$$

The EWS Lagrangian is part of the complete UST Lagrangian that includes the Generation groups, the Layer groups, and the ElectroWeak particle interactions. Each of the four layers of the Cosmos Theory UST has a different set of symmetry groups. Additional details may be found in chapter 6 of Blaha (2026b).

3. Extended General Relativity (EGR)[4]

The EGR Gravitation Lagrangian[5] is

$$\mathcal{L}_{PQ}' = \mathcal{L}_B - [(M^4(R^{1\mu\nu} + M^2 g^{1\mu\nu})(R^2_{\mu\nu} + AM^2 g^{2\mu\nu}) + ½M^4(R^{1\mu\nu} + M^2 g^{1\mu\nu})(R^1_{\mu\nu} + M^2 g^{1\mu\nu}) + \\ + \lambda_R(g^{1\mu\nu}g^1_{\mu\nu} - \eta^{\mu\nu}\eta_{\mu\nu})^2 + \lambda_{C1}g^{1\mu\nu}g^2_{\mu\nu} + \lambda_{C2}g^{1\mu\nu}g^1_{\mu\nu} + \lambda_{C3}g^{2\mu\nu}g^2_{\mu\nu} + M^4(g_G{}^2/4\pi)[M1]g^{1\mu\nu}J_{\mu\nu}] \\ + CC \tag{3.1}$$

from eqs. 4.1 and 4.38 of chapter 4 of Blaha (2026d) including the additional terms $\mathcal{L}_B$ for the abelian U(1) gauge fields: $B^{1\mu\nu}$ and $B^{2\mu\nu}$ that lead to an Absolute space-time:[6]

$$\mathcal{L}_B = -½B^{1\mu\nu}B^2_{\mu\nu} - ¼\, B^{1\mu\nu}B^1_{\mu\nu} - V_B(B^{1\mu\nu}) - b_2(B^{2\mu}g^1_{\mu\nu})^\dagger\, g^{1\nu\xi}B^2_\xi - \\ - b_1(B^{1\mu}g^1_{\mu\nu})^\dagger g^{1\nu\xi}B^1_\xi + g_B B^\mu J_\mu + CC \tag{3.2}$$

with a B condensate with potential

$$V_B(B^{1\mu}) = \lambda_B(B^{1\mu}B^1_\mu - B_0{}^\mu B_{0\mu})^2 \tag{3.3}$$

and

$$B_0{}^\mu = (0,\ 0,\ 0,\ b_0) \tag{3.4}$$

$$B^i_{\mu\nu} = \partial B^i_\mu/\partial x^\nu - \partial B^i_\nu/\partial x^\mu \tag{3.5}$$

Only the z component of $B_0{}^\mu$ is non-zero. A spatial rotation of the coordinate system makes all spatial components non-zero.

For the time-like ν = 0 coordinate we found a mass term for ε^0 from:

$$\square\varepsilon^\nu + 4b_1\varepsilon^\nu \tag{3.6}$$

which gives the time-like B^{10} mass:

$$m_{B0}{}^2 = 4b_1 \tag{3.7}$$

For ν = 3 (and all spatial coordinates after spatial rotations) we found a mass term for ε^3 :

$$\square\varepsilon^3 + (16\lambda_B b_0{}^2 + 4b_1)\varepsilon^3 + 3b_1 B_{0\xi}h^{1\nu\xi})\varepsilon^3 \tag{3.8}$$

---

[4] From Blaha (2026d).

[5] The √g factor that appears in Gravity Lagrangians has a higher order effect that is not relevant for leading order low gravitation approximations. We therefore focus on Lagrangian densities. When we refer to the Lagrangian we often mean Lagrangian density depending on the context.

[6] These fields are similar to the Electromagnetic gauge fields. We set the interaction term coupling constant g_B to zero in this chapter.

which gives a mass:

$$m_{Bspatial}^2 = 16\lambda_B b_0^2 + 4b_1 + 3b_1 b_0 \qquad (3.9)$$

The $B_0{}^{\nu} B_0{}^{\mu} \varepsilon^3$ term value is $b_0{}^2 \varepsilon^3$ and the $3b_1 B_{0\xi} h^{1\nu\xi}$ is only non-zero for $\nu = 3$.

Thus the $B^{1\mu}$ field masses breaks Lorentz covariance through the $B^{1\mu}$ condensate. The ε^0 mass squared is

$$m_{Bspatial}^2 = 16\lambda_B b_0^2 + 4b_1 + 3b_1 b_0 \qquad (3.10)$$

Note b_1 and λ_B both have the dimensions of [mass^2].

The condensate also was found to break Lorentz covariance of the metric tensors h^1 and h^2 and to establish an Absolute coordinate system – namely the rest frame of the B condensate.

Since

$$\Box B^{2\nu} = \tfrac{1}{4} \Box^2 \varepsilon^{\nu}/b_2 \qquad (3.11)$$

we found $B^{2\nu}$ has the masses of $B^{1\nu}$, and also a zero mass part, if $b_2 \neq 0$.

The discussion of other EGR features can be found in Blaha (2026d).

4. Enhanced Extended General Relativity (EEGR) by Analogy with EWS

There is a relationship in form between the ElectroWeakStrong (EWS) Lagrangian and a new enhancement of the Extended Gravitation (EGR) Lagrangian indicated in eqs. 3.1 and 3.2.[7] We will derive a New enhanced Gravitation below using a map from the EWS Lagrangian.[8] As a result we obtain a new set of Field Equations that have remarkable features such as an Absolute Reference Frame, the form of the Gravitation constant G, and support for a Theory of Time-Space. (See chapter 5)

The great benefit of a map from EWS is that it constrains the Gravitation theory and thus prevents the pointless introduction of features. It also suggests the Physical Reality of the Cosmos Theory, the UST, the EWS and the EEGR.

An especially important EEGR feature is the appearance of quartic terms in the field equations. These terms result in a finite[9] Gravitation Perturbation Theory based on a negative superficial degree of divergence for all EEGR perturbation theory diagrams. Negativity implies a convergent – non-infinite value for a diagram. The quartic terms generated finiteness corresponds to the EWS confinement of color fermions and gluons.

The map uses the EWS Lagrangian as a template and directly derives the new Enhanced Extended Gravitation (EEGR) Lagrangian below. *We will be mapping the ElectroWeak part of EWS to the B field part of the EEGR Lagrangian, and the Strong interactions part of EWS to the $R^{\mu\nu}$ parts of EEGR.* The map is based on the similarities of EWS SU(2)⊗U(1)⊗SU(4) with EEGR:

	EWS		
	Higgs Particles	SU(2)⊗U(1)	SU(4)⊗U(1)
Number of a Generators	4	4	16

	EEGR		
	Vector φ	B fields	$g^{\mu\nu}$
Number Lorentz indices	4	4	16

Our purpose is twofold: first to correlate Gravitation and Cosmos Symmetries as portrayed above relating ElectroWeak with the B field and the Strong interactions with the metric tensor dynamics. Secondly, we seek a possible further extension of Gravitation phenomena based on mapping aspects of the EWS theory.

---

[7] The connection between space-time and EWS symmetry groups was initially explored in Blaha (2018e).

[8] In 2018 we created a model EWS based on complex 4D space that contained the ElectroWeak and Strong interactions. See Blaha (2018e).

[9] See chapter 8.

The map deletes the space-time indices in the EWS Lagrangian and replaces internal symmetry indices with space-time indices in the below fashion:[10]

φ_{1b}	φ_{1sa}	$W_1{}^{\mu}{}_b$	$A_1{}^{\mu}$	$W_{1b}{}^{\mu\nu}$	$F_{1a}{}^{\mu\nu}$	$F_{2a}{}^{\mu\nu}$	(4.1)
$\varphi^1{}_\beta$ For the B fields	$\varphi^1{}_{sa}$ For the $g^{\mu\nu}$ fields	$B^{1\beta}$	$g^{1\alpha\gamma}$	$B^{1\beta\gamma}$	$R'^{1\alpha\gamma}$	$R'^{2\alpha\gamma}$	

and

φ_{2b}	φ_{2sa}	$W_2{}^{\mu}{}_b$	$A_2{}^{\mu}$	$W_{2b}{}^{\mu\nu}$	$F_{2a}{}^{\mu\nu}$	$F_{1a}{}^{\mu\nu}$
$\varphi^2{}_\beta$ For the B fields	$\varphi^2{}_{sa}$ For the $g^{\mu\nu}$ fields	$B^{2\beta}$	$g^{2\alpha\gamma}$	$B^{2\beta\gamma}$	$R'^{2\alpha\gamma}$	$R'^{1\alpha\gamma}$

where the b index and the β index run from 1 through 4, and the a index and the β and γ indices also run from 1 through 4; and

$$R'^{1\mu\nu} = M^2 R^{1\mu\nu} + M^4 g^{1\mu\nu}$$
$$R'^{2\mu\nu} = M^2 R^2{}_{\mu\nu} + AM^4 g^{2\mu\nu} \tag{4.2}$$

where A is a constant that we will take to have the value 0. The mass factors lead to the [dimension] of the Gravitation constant G. *The potential for the metric tensor field $g^{1\mu\nu}$ sets its minimum value to the usual weak field approximation $\eta^{\mu\nu}$ for gravitational radiation.*

We have chosen to define the EEGR Higgs fields as vector fields to enforce the map of symmetry indices to Lorentz indices. We now assume Higgs fields and the B fields are real-valued and implement the map:

FROM:

$$\begin{aligned}\mathcal{L}_{EWS} = {} & (\partial_\mu \varphi_{1b}{}^*)(\partial^\mu \varphi_{2b}) + (\partial_\mu\, \varphi_{1sa}{}^*)(\partial^\mu \varphi_{2sa}) \; - \tfrac{1}{2} W_{1b}{}^{\mu\nu} W_{2b\mu\nu} - \tfrac{1}{2} F_{1a}{}^{\mu\nu} F_{2a\mu\nu} - \\ & - (W_2{}^\mu \varphi_1)_b{}^\dagger (W_{2\mu}\varphi_1)_b - (W_1{}^\mu \varphi_2)_b{}^\dagger (W_{1\mu}\varphi_2)_b - (A_1{}^\mu\, \varphi_{1s})_a{}^\dagger (A_1{}^a{}_\mu \varphi_{1s})_a - (A_2{}^\mu \varphi_{2s})_a{}^\dagger (A_{2\mu}\varphi_{2s})_a - \\ & - V(\varphi_1) - m_1{}^2 \varphi_{1b}{}^* \varphi_{2b} - \tfrac{1}{2} m_2{}^2 \varphi_{1b}{}^* \varphi_{1b} - \tfrac{1}{2}\, m_3{}^2 \varphi_{2b}{}^* \varphi_{2b} - \\ & - V(\varphi_{1s}) - m_{1s}{}^2 \varphi_{1sa}{}^* \varphi_{2sa} - \tfrac{1}{2} m_{2s}{}^2 \varphi_{1sa}{}^* \varphi_{1sa} - \tfrac{1}{2}\, m_{3s}{}^2 \varphi_{2sa}{}^* \varphi_{2sa} + CC\end{aligned} \tag{2.2}$$

TO
the corresponding EWSGR Lagrangian terms:

$$\begin{aligned}\mathcal{L}_{EWSGR} = {} & F^{\mu\nu}(\varphi^1) F_{\mu\nu}(\varphi^2) + F^{\mu\nu}(\varphi^{1s}) F_{\mu\nu}(\varphi^{2s}) \; - \tfrac{1}{2} B^{1\mu\nu} B^2{}_{\mu\nu} - \tfrac{1}{2}\, R'^{1\alpha\gamma}\, R'^2{}_{\alpha\gamma} - \\ & - B_2{}^\mu \varphi^{1\nu} B^2{}_\mu \varphi^1{}_\nu - B^{1\mu} \varphi^{2\nu} B^1{}_\mu \varphi^2{}_\nu - g^{1\mu\nu}\, \varphi^1{}_s{}^\alpha g^1{}_{\mu\nu} \varphi^1{}_{s\alpha} - g^{2\mu\nu} \varphi^2{}_s{}^\alpha g^2{}_{\mu\nu} \varphi^2{}_{s\alpha} - \\ & - V(\varphi^1) - m_1{}^2 \varphi^{1\mu} \varphi^2{}_\mu - \tfrac{1}{2} m_2{}^2 \varphi^{1\mu} \varphi^1{}_\mu - \tfrac{1}{2}\, m_3{}^2\, \varphi^{2\mu} \varphi^2{}_\mu - \\ & - V(\varphi^1{}_s) - - m_{1s}{}^2 \varphi^1{}_s{}^\mu \varphi^2{}_{s\mu} - \tfrac{1}{2} m_{2s}{}^2 \varphi^1{}_s{}^\mu \varphi^1{}_{s\mu} - \tfrac{1}{2}\, m_{3s}{}^2 \varphi^2{}_s{}^\mu \varphi^2{}_{s\mu} \; + CC\end{aligned} \tag{4.3}$$

---

[10] Note also the change from lower index to upper index PseudoQuantum indices.

after raising the PseudoQuantum indices. *Note we made the Higgs particles into vectors to conform with the map from symmetry indices to Lorentz indices.*

Now by adding additional EGR terms for the metric field, and mapping mass terms to Cosmological constants we find

$$\mathscr{L}_{EEGR} = \mathscr{L}_{EWSGR} + \mathscr{L}_{PQ}'' + \mathscr{L}_{EB} \tag{4.4}$$

where

$$F^{\mu\nu}(\varphi) = \partial^\mu\varphi^\alpha\partial^\nu\varphi_\alpha - \partial^\nu\varphi^\alpha\partial^\mu\varphi_\alpha$$

and

$$\begin{aligned}\mathscr{L}_{PQ}'' = -\,[&\tfrac{1}{2} R'^{1\alpha\gamma} R'^{2}{}_{\alpha\gamma} + \tfrac{1}{2} R'^{1\alpha\gamma} R'^{1}{}_{\alpha\gamma} + V_g(g^{1\mu\nu}) + \lambda_{C1}g^{1\mu\nu}g^{2}{}_{\mu\nu} + \lambda_{C2}g^{1\mu\nu}g^{1}{}_{\mu\nu} + \\ &+ \lambda_{C3}g^{2\mu\nu}g^{2}{}_{\mu\nu} + M^4(g_G{}^2/4\pi)[M2]g^{1\mu\nu}J_{\mu\nu}] + CC\end{aligned} \tag{4.5}$$

and

$$\begin{aligned}\mathscr{L}_{EB} = &-\tfrac{1}{2}B^{1\mu\nu}B^{2}{}_{\mu\nu} - \tfrac{1}{4}B^{1\mu\nu}B^{1}{}_{\mu\nu} - V_B(B^{1\mu\nu}) - b_2(B^{2\mu}g^{1}{}_{\mu\nu})^\dagger\, g^{1\nu\xi}B^{2}{}_\xi - \\ &- b_1(B^{1\mu}g^{1}{}_{\mu\nu})^\dagger g^{1\nu\xi}B^{1}{}_\xi + g_B B^{1\mu}J_\mu + CC\end{aligned} \tag{4.6}$$

Comparing eqs. 4.4 – 4.6 to the equations and solutions in chapter 5 of Blaha (2026d) we find this map generates the *extra* Lagrangian density terms:[11]

$$\begin{aligned}\mathscr{L}_{extra} = &F^{\mu\nu}(\varphi^1)F_{\mu\nu}(\varphi^2) + F^{\mu\nu}(\varphi^{1s})F_{\mu\nu}(\varphi^{2s}) - B^{2\mu}\varphi^{1\nu}B^{2}{}_\mu\varphi^{1}{}_\nu - B^{1\mu}\varphi^{2\nu}B^{1}{}_\mu\varphi^{2}{}_\nu - g^{1\mu\nu}\varphi^{1}{}_s{}^\alpha g^{1}{}_{\mu\nu}\varphi^{1}{}_{s\alpha} - \\ &- g^{2\mu\nu}\varphi^{2}{}_s{}^\alpha g^{2}{}_{\mu\nu}\varphi^{2}{}_{s\alpha} - V(\varphi^1) - m_1{}^2\varphi^{1\mu}\varphi^{2}{}_\mu - \tfrac{1}{2}m_2{}^2\varphi^{1\mu}\varphi^{1}{}_\mu - \tfrac{1}{2}\, m_3{}^2\, \varphi^{2\mu}\varphi^{2}{}_\mu - \\ &- V(\varphi^{1}{}_s) - - m_{1s}{}^2\varphi^{1}{}_s{}^\mu\varphi^{2}{}_{s\mu} - \tfrac{1}{2}m_{2s}{}^2\varphi^{1}{}_s{}^\mu\varphi^{1}{}_{s\mu} - \tfrac{1}{2}\, m_{3s}{}^2\varphi^{2}{}_s{}^\mu\varphi^{2}{}_{s\mu} + CC\end{aligned} \tag{4.7}$$

where the potentials are

Purpose

$$V(\varphi^1) = \sum_{a=1}^{4} \lambda_{\varphi a}(\varphi^{1}{}_a{}^*\varphi^{1}{}_a - \varphi^{1}{}_{0a}{}^2)^2 \tag{4.8}$$

set the ElectroWeak Higgs field minimum

$$V(\varphi^{1}{}_s) = \sum_{a=1}^{16} \lambda_{\varphi sa}(\varphi^{1}{}_{sa}{}^*\varphi^{1}{}_{sa} - \varphi^{1}{}_{s0a}{}^2)^2 \tag{4.9}$$

set the SU(4) Higgs field minimum

$$V_B(B^{1\mu\nu}) = \lambda_B(B^{1\mu}B^{1}{}_\mu - B_0{}^\mu B_{0\mu})^2 \tag{4.10}$$

set the B field minimum

$$V_g(g^{1\mu\nu}) = \lambda_R(g^{1\mu\nu}g^{1}{}_{\mu\nu} - \eta^{\mu\nu}\eta_{\mu\nu})^2 \tag{4.11}$$

set the metric tensor minimum

The minimum for $V(\varphi^1)$ is set by

$$\begin{aligned}\varphi^{1}{}_0{}^0 &= 0 &&\text{for } i = 0 \\ \varphi^{1}{}_0{}^i &= \varphi^{1}{}_0 &&\text{for } i = 2, 3, 4\end{aligned} \tag{4.12}$$

---

[11] Note the terms in eqs. 4.3 and 4.5 do not overlap.

The minimum for $V(\varphi^1{}_s)$ is set by

$$\varphi^1{}_{s0}{}^0 = 0 \qquad \text{for } i = 0 \tag{4.13}$$
$$\varphi^1{}_{s0}{}^i = \varphi^1{}_{s0} \qquad \text{for } i = 2, 3, \ldots, 16$$

The $\mathscr{L}_{EEGR}$ Lagrangian density leads to four additional Lagrangian variations for the Higgsian fields with consequent dynamic equations, and to additional terms in the B fields and $h^{\mu\nu\,metric}$ fields Lagrangian variations and dynamic equations.

4.1 Higgs Fields Lagrangian Variation for φ^{iv}

We choose the gauge conditions

$$\partial^\mu\varphi^i{}_\mu = \partial^\mu\varphi^i{}_{s\mu} = 0$$

for i = 1, 2.

The φ^{1v} Higgs Fields Lagrangian variation leads to

$$0 = \Box\varphi^{2v} + 2\lambda_\varphi(\varphi^{1\mu *}\varphi^1{}_\mu - \varphi^1{}_{0\mu}{}^2)\varphi^{1v} + B^{2\mu}B^2{}_\mu\varphi^{1v} + m_1{}^2\varphi^{2v} + m_2{}^2\varphi^{1v} \tag{4.14}$$
$$= \Box\varphi^{2v} + 2\lambda_\varphi(\varphi^{1\mu *}\varphi^1{}_\mu - \varphi^1{}_{0\mu}{}^2)\varphi^{1v} + m_1{}^2\varphi^{2v} + m_2{}^2\varphi^{1v}$$

to leading order where we suppress the summation in the potential.

The φ^{2v} Higgs Fields Lagrangian variation leads to

$$0 = \Box\,\varphi^{1v} + m_1{}^2\varphi^{1v} + m_3{}^2\varphi^{2v} + B^{1\mu}B^1{}_\mu\varphi^{2v} \tag{4.15}$$
$$= \Box\,\varphi^{1v} + m_1{}^2\varphi^{1v} + m_3{}^2\varphi^{2v} + B_0{}^\mu B_{0\mu}\varphi^{2v} \tag{4.16}$$

to leading order. Eq. 4.14 can be put in the form

$$0 = \Box\,\varphi^{1v} + m_1{}^2\varphi^{1v} + m_3'^2\varphi^{2v} \tag{4.17}$$

where

$$m_3'^2 = m_3{}^2 + B_0{}^\mu B_{0\mu}$$

Then we may use the results of chapter 9 of Blaha (2026b) by using m_3' and the vector equivalent of its scalar formulation. The dynamic eqs. 4.16 and 4.17 then have their form. Thus from chapter 9 we see:

We set

$$\varphi^{1v}(x) = \varphi^1{}_0{}^v + \varepsilon^v(x) \tag{9.5}$$

where ε v (x) is small. Then eq. [4.16] becomes approximately

$$(\Box + m_1{}^2)\varepsilon^v + m_1{}^2\varphi^1{}_0{}^v + m_3'^2\varphi^{2v} = 0 \tag{9.6}$$

and eq. [4.14] gives (surpressing the "a" dependence in eq. 4.8)

$$(\square + m_1^2)\varphi^{2v} + 4\lambda_{\varphi}\varphi^1{}_0{}^2\,\varepsilon^v(\delta^{v2} + \delta^{v3} + \delta^{v4}) + m_2^2\varepsilon^v + m_2^2\varphi^1{}_0{}^v = 0 \quad (9.7)$$

to leading order in ε. [We have inserted the delta function factor due to the vector nature of this Higgs field. It results in differences between the time component mass and the spatial component mass. This breakdown of Lorentz symmetry (Absolute reference frame) corresponds directly to symmetry breaking in the particle sector.]

Multiplying by $(\square + m_1^2)$ we find

$$(\square + m_1^2)^2\varphi^{2v} - (4\lambda\,\varphi^1{}_0{}^{v2} + m_2^2)m_3'^2\varphi^{2v} - (4\lambda_{\varphi}\varphi^1{}_0{}^2\,\varepsilon^v(\delta^{v2} + \delta^{v3} + \delta^{v4}) + m_2^2)m_1^2\varphi^1{}_0{}^v + m_1^2m_2^2\varphi^1{}_0{}^v = 0 \quad (9.8)$$

or

$$[(\square + m_1^2)^2 - (4\lambda_{\varphi}\varphi^1{}_0{}^2\,\varepsilon^v(\delta^{v2} + \delta^{v3} + \delta^{v4}) + m_2^2)m_3'^2]\varphi^{2v} - (4\lambda\,\varphi^1{}_0{}^2 + m_2^2)m_1^2\varphi^1{}_0 + m_1^2m_2^2\varphi^1{}_0 = 0 \quad (9.9)$$

Eq. 9.9 implies the φ^2 field propagator has the masses

$$M_2^{v2} = m_1^2 + [(4\lambda_{\varphi}\varphi^1{}_0{}^2\,\varepsilon^v(\delta^{v2} + \delta^{v3} + \delta^{v4}) + m_2^2)\,m_3'^2]^{1/2} \quad (9.10a)$$

or

$$M_2^{v2} = m_1^2 - [(4\lambda_{\varphi}\varphi^1{}_0{}^2\,\varepsilon^v(\delta^{v2} + \delta^{v3} + \delta^{v4}) + m_2^2)\,m_3'^2]^{1/2} \quad (9.10b)$$

and differ for the time-like v = 0 vs. the space-like mass v = 2, 3, 4.

Eq. 9.6 implies

$$[(\square + m_1^2)^2 - (4\lambda_{\varphi}\varphi^1{}_0{}^2\varepsilon^v(\delta^{v2} + \delta^{v3} + \delta^{v4}) + m_2^2)m_3'^2](\square + m_1^2)\varepsilon_1 + m_1^6\varphi^1{}_0 + m_3^2(4\lambda\,\varphi^1{}_0{}^2 + m_2^2)m_1^2m_2^2\varphi^1{}_0 = 0 \quad (9.11)$$

The φ^1 field propagator (Higgs ε field propagator) has masses that also differ for the time-like and spatial cases:

$$M_1^{v2} = m_1^2 \quad (9.12a)$$
$$M_1^{v2} = m_1^2 + [(4\lambda_{\varphi}\varphi^1{}_0{}^2\,\varepsilon^v(\delta^{v2} + \delta^{v3} + \delta^{v4}) + m_2^2)\,m_3'^2]^{1/2} \quad (9.12b)$$
$$M_1^{v2} = m_1^2 - [(4\lambda_{\varphi}\varphi^1{}_0{}^2\,\varepsilon^v(\delta^{v2} + \delta^{v3} + \delta^{v4}) + m_2^2)\,m_3'^2]^{1/2} \quad (9.12c)$$

The Higgs Mechanism results from the non-zero vacuum expectation value $\varphi^1{}_0$ for the φ^1 field.

Note the masses depend on the Lorentz index of the Higgs vector and are part of the breaking of Lorentz symmetry resulting in an Absolute reference frame – directly comparable to "normal" Higgsian symmetry breaking.

4.2 Higgs Fields Lagrangian Variation for $\varphi^i{}_s{}^v$

We now consider the Higgs vector fields corresponding to the $h^{i\mu v}$ fields (the analogues of the Strong A gauge fields) for PseudoQuantum i = 1, 2.

The $\varphi^1{}_s{}^v$ Higgs Fields Lagrangian variation gives (surpressing the "a" dependence in eq. 4.9)

$$0 = \square\varphi^2{}_s{}^v + 2\lambda_{\varphi s}(\varphi^1{}_s{}^{\mu *}\varphi^1{}_\mu - \varphi^1{}_{s0\mu}{}^2)\varphi^1{}_s{}^v + m_{1s}{}^2\varphi^2{}_s{}^v + 2m_{2s}{}^2\varphi^1{}_s{}^v + g^{1\mu\beta}\varphi^1{}_s{}^v g^1{}_{\mu\beta} \quad (4.18)$$
$$= \square\varphi^2{}_s{}^v + 2\lambda_{\varphi s}(\varphi^1{}_s{}^{\mu *}\varphi^1{}_{s\mu} - \varphi^1{}_{s0\mu}{}^{2s})\varphi^1{}_s{}^v + m_{1s}{}^2\varphi^2{}_s{}^v + 2m_{2s}{}^2\varphi^1{}_s{}^v + 8\,\varphi^1{}_s{}^v \quad (4.19)$$

at the minimum of the $V_g(g^{1\mu\nu})$ potential. We now set

$$\varphi^{1\,\nu}_{s}(x) = \varphi^{1}_{s0}{}^{\nu} + \varepsilon_s{}^{\nu}(x) \tag{4.20}$$

where $\varepsilon_s{}^{\nu}(x)$ is small. Then eq. 4.19 becomes approximately

$$(\Box + m_{1s}{}^2)\varphi^{2\,\nu}_{s} + (4\lambda_{\varphi s}\varphi^{1}_{s}{}^{0\,2}\Sigma\delta^{\nu i} + 2m_{2s}{}^2 + 8)\varepsilon_s{}^{\nu} + (2m_{2s}{}^2 + 8)\varphi^{1}_{s0} = 0 \tag{4.21}$$

where the sum extends from 2 through 16 due to eq. 4.13.

The $\varphi^{2\,\nu}_{s}$ Higgs Fields Lagrangian variation gives (suppressing the "a" dependence in the potential)

$$0 = \Box\,\varphi^{1\,\nu}_{s} + m_{1s}{}^2\varphi^{1\,\nu}_{s} + 2m_{3s}{}^2\varphi^{2\,\nu}_{s} + 8\,\varphi^{2\,\nu}_{s} \tag{4.22}$$

to leading order.

Eq. 4.22 can be put in the form

$$0 = (\Box + m_{1s}{}^2)\varepsilon_s{}^{\nu} + m_{1s}{}^2\varphi^{1}_{s0} + (2m_{3s}{}^2 + 8)\varphi^{2\,\nu}_{s} \tag{4.23}$$

at the minimum of the $V_g(g^{1\mu\nu})$ potential.

Applying $(\Box + m_{1s}{}^2)$ to eq. 4.21 and substituting from eq. 4.23 we obtain

$$(\Box + m_{1s}{}^2)^2\varphi^{2\,\nu}_{s} - (4\lambda_{\varphi s}\varphi^{1}_{s}{}^{0\,2}\Sigma\delta^{\nu i} + 2m_{2s}{}^2 + 8)[\,m_{1s}{}^2\varphi^{1}_{s0} + (2m_{3s}{}^2 + 8)\varphi^{2\,\nu}_{s}] + m_{1s}{}^2(2m_{2s}{}^2 + \\ + 8)\varphi^{1}_{s0} = 0$$

$$(\Box + m_{1s}{}^2)^2\varphi^{2\,\nu}_{s} - (4\lambda_{\varphi s}\varphi^{1}_{s}{}^{0\,2}\Sigma\delta^{\nu i} + 2m_{2s}{}^2 + 8)(2m_{3s}{}^2 + 8)\varphi^{2\,\nu}_{s} - \\ - (4\lambda_{\varphi s}\varphi^{1}_{s}{}^{0\,2}\Sigma\delta^{\nu i} + 2m_{2s}{}^2 + 8)\,m_{1s}{}^2\varphi^{1}_{s0} + m_{1s}{}^2(2m_{2s}{}^2 + 8)\varphi^{1}_{s0} = 0$$

or

$$[(\Box + m_{1s}{}^2)^2 - (4\lambda_{\varphi s}\varphi^{1}_{s}{}^{0\,2}\Sigma\delta^{\nu i} + 2m_{2s}{}^2 + 8)(2m_{3s}{}^2 + 8)]\varphi^{2\,\nu}_{s} - \\ - 4\lambda_{\varphi s}\varphi^{1}_{s}{}^{0\,2}\Sigma\delta^{\nu}m_{1s}{}^2\varphi^{1}_{s0} + m_{1s}{}^2(2m_{2s}{}^2 + 8)\varphi^{1}_{s0} = 0 \tag{4.24}$$

Eq. 4.24 implies the $\varphi^{2\,\nu}_{s}$ field propagators have the masses

$$M_{2s}{}^{\nu\,2} = m_{1s}{}^2 + [(4\lambda_{\varphi s}\varphi^{1}_{s}{}^{0\,2}\Sigma\delta^{\nu i} + 2m_{2s}{}^2 + 8)(2m_{3s}{}^2 + 8)]^{1/2} \tag{4.25}$$

or

$$M_{2}{}^{\nu\,2} = m_{1}{}^2 - [(4\lambda_{\varphi s}\varphi^{1}_{s}{}^{0\,2}\Sigma\delta^{\nu i} + 2m_{2s}{}^2 + 8)(2m_{3s}{}^2 + 8)]^{1/2} \tag{4.26}$$

and differ for the $\nu = 1$ index vs. the $\nu = 2, 3, \ldots, 16$ indices.

For $\nu = 1$ the masses are

$$M_{2s}{}^{1\,2} = m_{1s}{}^2 + [2m_{2s}{}^2 + 8)(2m_{3s}{}^2 + 8)]^{1/2} \tag{4.27}$$

or

$$M_{2s}{}^{1\,2} = m_{1s}{}^2 - [2m_{2s}{}^2 + 8)(2m_{3s}{}^2 + 8)]^{1/2} \tag{4.28}$$

For $\nu = 2, 3, 4, .., 16$ the masses are

$$M_{2s}{}^{1a\,2} = m_{1s}{}^2 + [4\lambda_{\varphi s}\varphi^1{}_s{}^{0\,2} + 2m_{2s}{}^2 + 8)(2m_{3s}{}^2 + 8)]^{1/2} \quad (4.27a)$$

or

$$M_{2s}{}^{1a\,2} = m_{1s}{}^2 - [4\lambda_{\varphi s}\varphi^1{}_s{}^{0\,2} + 2m_{2s}{}^2 + 8)(2m_{3s}{}^2 + 8)]^{1/2} \quad (4.28a)$$

The ε masses may also be similarly determined from eqs. 4.21 and 4.23.

4.3 The B Fields $B^{1\mu\nu}$ and $B^{2\mu\nu}$ Variations

We showed the B fields, $B^{1\mu\nu}$ and $B^{2\mu\nu}$, lead to an Absolute space-time in Blaha (2026d). We now add the terms $-B^{2\mu}\varphi^{1\nu}B^2{}_\mu\varphi^1{}_\nu - B^{1\mu}\varphi^{2\nu}B^1{}_\mu\varphi^2{}_\nu$ that appear in $\mathscr{L}_{extra}$. The relevant Lagrangians are

$$\mathscr{L}_{EB} = -\tfrac{1}{2}B^{1\mu\nu}B^2{}_{\mu\nu} - \tfrac{1}{4}B^{1\mu\nu}B^1{}_{\mu\nu} - V_B(B^{1\mu\nu}) - b_2(B^{2\mu}g^1{}_{\mu\nu})^\dagger g^{1\nu\xi}B^2{}_\xi - \\ - b_1(B^{1\mu}g^1{}_{\mu\nu})^\dagger g^{1\nu\xi}B^1{}_\xi + g_B B^{1\mu}J_\mu + CC \quad (4.6)$$

and

$$\mathscr{L}_{extra} = F^{\mu\nu}(\varphi^1)F_{\mu\nu}(\varphi^2) + F^{\mu\nu}(\varphi^{1s})F_{\mu\nu}(\varphi^{2s}) - B^{2\mu}\varphi^{1\nu}B^2{}_\mu\varphi^1{}_\nu - B^{1\mu}\varphi^{2\nu}B^1{}_\mu\varphi^2{}_\nu - \\ - g^{1\mu\nu}\varphi^1{}_s{}^\alpha g^1{}_{\mu\nu}\varphi^1{}_{s\alpha} - g^{2\mu\nu}\varphi^2{}_s{}^\alpha g^2{}_{\mu\nu}\varphi^2{}_{s\alpha} - V(\varphi^1) - m_1{}^2\varphi^{1\mu}\varphi^2{}_\mu - \tfrac{1}{2}m_2{}^2\varphi^{1\mu}\varphi^1{}_\mu - \tfrac{1}{2}m_3{}^2\varphi^{2\mu}\varphi^2{}_\mu - \\ - V(\varphi^1{}_s) - m_{1s}{}^2\varphi^1{}_s{}^\mu\varphi^2{}_{s\mu} - \tfrac{1}{2}m_{2s}{}^2\varphi^1{}_s{}^\mu\varphi^1{}_{s\mu} - \tfrac{1}{2}m_{3s}{}^2\varphi^2{}_s{}^\mu\varphi^2{}_{s\mu} + CC \quad (4.7)$$

where we have a B condensate with potential

$$V_B(B^{1\mu}) = \lambda_B(B^{1\mu}B^1{}_\mu - B_0{}^\mu B_{0\mu})^2 \quad (4.7a)$$

and

$$B_0{}^\mu = (0,\ 0,\ 0,\ b_0) \\ B^i{}_{\mu\nu} = \partial B^i{}_\mu/\partial x^\nu - \partial B^i{}_\nu/\partial x^\mu \quad (4.7b)$$

Only the z component of $B_0{}^\mu$ is non-zero. Later we will see the spatial rotation of the coordinate system implies all spatial components are non-zero.

From the variation in $B^{2\mu}$ in $\mathscr{L}_B$ and the new terms we obtain the dynamic equation:

$$\partial_\mu B^{1\mu\nu} + 4\,b_2 g^{1\mu\nu}g^1{}_\mu{}^\xi B^2{}_\xi + 2\varphi^1{}_\alpha\varphi^{1\alpha}B^{2\nu} = 0 \quad (4.29)$$

From the Lagrangian's variation in $B^{1\mu}$ we obtain the dynamic equation

$$\tfrac{1}{2}\,\partial_\mu B^{2\mu\nu} + \tfrac{1}{2}\,\partial_\mu B^{1\mu\nu} + \partial/(\partial B^1{}_\nu)V_B(B^{1a\mu\nu}) + 2\,b_1 g^1{}_\mu{}^\nu g^{1\mu\xi}B^1{}_\xi + 2\varphi^2{}_\alpha\varphi^{2\alpha}B^{1\nu} = 0 \quad (4.30)$$

Defining an increment field $\varepsilon_B{}^\mu$ where

$$B^{1\mu} = B_0{}^\mu + \varepsilon_B{}^\mu \quad (4.31)$$

We find eq. 4.30 becomes

$$\partial_\mu B^{2\mu\nu} + \partial_\mu B^{1\mu\nu} + 16\lambda_B B_0{}^\nu B_0{}^\mu\varepsilon_{B\mu} + 4(b_1\eta_\mu{}^\nu\eta^{\mu\xi}\varepsilon_{B\xi} + b_1\eta_\mu{}^\nu h^{1\mu\xi}B_{0\xi} + \\ + b_1 h^1{}_\mu{}^\nu\eta^{1\mu\xi}B_{0\xi} + b_1\eta_\mu{}^\nu\eta^{1\mu\xi}B_{0\xi}) = 0 \quad (4.32)$$

after discarding nonlinear terms. Choosing the gauges:

$$\partial_\mu B^{1\mu} = \partial_\mu \varepsilon_B{}^\mu = 0 \quad (4.33)$$
$$\partial_\mu B^{2\mu} = 0 \quad (4.34)$$

eq. 4.32 becomes

$$\Box\varepsilon_B{}^\nu + \Box B^{2\nu} + 16\lambda_B B_0{}^\nu B_0{}^\mu \varepsilon_{B\mu} + 4b_1\varepsilon_B{}^\nu + 12b_1 B_0{}^\nu = 0 \quad (4.35)$$

to leading order after discarding nonlinear terms and using the minimum values for $h^{1\nu\xi}$.

Eq. 4.29 becomes

$$\Box\varepsilon_B{}^\nu + 4\, b_2 g^1{}_\mu{}^\nu g^{1\mu\xi} B^2{}_\xi + 2\varphi^1{}_0{}^2 B^{2\nu} = 0 \quad (4.36)$$

or

$$\Box\varepsilon_B{}^\nu + 4\, b_2 B^{2\nu} + 2\varphi^1{}_0{}^2 B^{2\nu} = 0 \quad (4.37)$$

to leading order where $\Box = \partial_\mu\partial^\mu$.

Applying $\Box$ to eq. 4.37 we obtain

$$\Box^2\varepsilon_B{}^\nu + (4b_2 + 2\varphi^1{}_0{}^2)\Box B^{2\nu} = 0 \quad (4.38)$$

Then substituting in eq. 4.35 we see

$$\Box\varepsilon_B{}^\nu - (4\, b_2 + 2\varphi^1{}_0{}^2)^{-1}\Box^2\varepsilon_B{}^\nu + 16\lambda_B B_0{}^\nu B_0{}^\mu \varepsilon_{B\mu} + 4b_1\varepsilon_B{}^\nu + 12b_1 B_0{}^\nu = 0 \quad (4.39)$$

or

$$\Box\varepsilon_B{}^\nu + 16\lambda_B B_0{}^\nu B_0{}^\mu \varepsilon_{B\mu} + 4b_1\varepsilon_B{}^\nu + 12b_1 B_0{}^\nu = 0 \quad (4.40)$$

after neglecting the higher derivative term.

From eq. 4.40 we see that there are two masses: one for $\nu = 0$ – a time-like $\varepsilon_B{}^0$ field:

$$M_{\varepsilon B0} = 4b_1 \quad (4.41)$$

and a spatial $\varepsilon_B{}^{0i}$ field mass for coordinates i = 1, 2, 3:

$$M_{\varepsilon Bi} = 16\lambda_B b_0{}^2 + 4b_1 \quad (4.42)$$

These masses are an analogue of the W gauge fields Higgs breakdown massesif we set $b_1 = 0$.

Note the $\varphi^1{}_0{}^2 B^{2\nu}$ *term gives mass in the Higgs Mechanism approach as opposed to the explicit mass dependence appearing in the B fields discussion of Blaha (2026d). However the source of the spatial mass is the B condensate.*

Note b_1 and λ_B both have the dimensions of [mass^2].

The B condensate also breaks Lorentz covariance of the metric tensors h^1 and h^2 and establishes an Absolute coordinate system – namely the rest frame of the B condensate. We discuss this topic in more detail next.

The possible existence of B^1 and B^2 quanta with non-zero masses raises the possibility of concentrations of these quanta bound by gravitation. It also raises the possibility of their interaction with the metric tensors $g^1{}_\mu{}^\nu$ and $g^{2\mu\xi}$.

4.4 Metric $h^{i\mu\nu}$ Fields Variations

The extra Lagrangian terms in $\mathcal{L}_{extra}$ that are introduced by the addition of the vector Higgs fields $\varphi^1{}_s{}^\alpha$ are

$$- g^{1\mu\nu}\varphi^1{}_s{}^\alpha g^1{}_{\mu\nu}\varphi^1{}_{s\alpha} - g^{2\mu\nu}\varphi^2{}_s{}^\alpha g^2{}_{\mu\nu}\varphi^2{}_{s\alpha} \tag{4.43}$$

The PseudoQuantum metric fields can be expanded using the relations

$$g^1{}_{\mu\nu}(x) = \eta_{\mu\nu} + h^1{}_{\mu\nu}(x) \tag{4.44}$$
$$g^2{}_{\mu\nu}(x) = h^2{}_{\mu\nu}(x) \tag{4.45}$$

The terms

$$g^{1\mu\nu}\varphi^1{}_s{}^\alpha g^1{}_{\mu\nu}\varphi^1{}_{s\alpha} + g^{2\mu\nu}\varphi^2{}_s{}^\alpha g^2{}_{\mu\nu}\varphi^2{}_{s\alpha}$$

become

$$(\eta^{\mu\nu} + h^{1\mu\nu})(\eta_{\mu\nu} + h^1{}_{\mu\nu})\varphi^1{}_s{}^\alpha\varphi^1{}_{s\alpha} + h^{2\mu\nu}h^2{}_{\mu\nu}\varphi^2{}_s{}^\alpha\varphi^2{}_{s\alpha}$$

$$(4 + h^{1\mu\nu}h^1{}_{\mu\nu})(1 - \delta^{\nu 0})\varphi^1{}_0{}^2 + h^{2\mu\nu}h^2{}_{\mu\nu}\varphi^2{}_s{}^\alpha\varphi^2{}_{s\alpha} \tag{4.46}$$

using

$$h = h^\sigma{}_\sigma = 0 \tag{4.47}$$

We now turn to the dynamic equations.

4.4.1 The $h^{1\mu\nu}$ Variation Dynamic Equation

This variation began with the terms in eq. 4.27 of Blaha (2026d):[12]

$$0 = (\tfrac{1}{4}M^4\Box^2 + \tfrac{1}{2}(A-1)M^6\Box + (\lambda_{C1} - AM^8))h^{2\varphi\chi} + \tfrac{1}{2}M^4\Box^2 h^{1\varphi\chi} - M^6\Box h^{1\varphi\chi} + + (\lambda_{C2} - 2M^8)h^{1\varphi\chi} + (\lambda_{C2} - 2M^8)\eta^{\varphi\chi} - M^4(g_G{}^2/4\pi)J^{\varphi\chi} \qquad \text{2026d-(4.27)}$$

to leading order. Then we add the new terms:

$$I^{\varphi\chi} = - b_1B^{1\varphi}B^{1\chi} - b_2B^{2\varphi}B^{2\chi} - b_1(B_0{}^\varphi + \varepsilon_1{}^\varphi)(B_{0a} + \varepsilon_{1a})h^{1\chi a} - b_2B^{2\varphi}B^2{}_a h^{1\chi a} \qquad \text{2026d-(5.27)}$$

To leading order we found the leading linear term is

$$K^{\varphi\chi} \cong - b_1B_0{}^\varphi B_{0a}h^{1\chi a} \qquad \text{2026d-(5.28)}$$

Since the only non-zero component of $B_0{}^\varphi$ is the 3 component we found the only non-zero component of $K^{\varphi\chi}$ was

$$K^{33} \cong - b_1B_0{}^3B_{03}h^{33} = -b_1b_0{}^2h^{133} \qquad \text{2026d-(5.29)}$$

[12] We add a prefix 2026d to the equation numbers indicating they appear in Blaha (2026d).

or

$$K^{\varphi\chi} = -b_1 b_0^2 \delta_{\varphi 3}\delta_{\chi 3} h^{1\varphi\chi} \qquad \text{2026d-(5.30)}$$

Thus eq. 2026d-(4.27) above became (with the new addition from eq. 4.26 above):

$$0 = (\tfrac{1}{4}M^4\Box^2 + \tfrac{1}{2}(A-1)M^6\Box + (\lambda_{C1} - AM^8))h^{2\varphi\chi} + \tfrac{1}{2} M^4\Box^2 h^{1\varphi\chi} - M^6\Box h^{1\varphi\chi} + \\ + (\lambda_{C2} - 2M^8)h^{1\varphi\chi} + (\lambda_{C2} - 2M^8)\eta^{\varphi\chi} - M^4(g_G^2/4\pi)J^{\varphi\chi} + K^{\varphi\chi} + 2h^{1\varphi\chi}(1 - \delta^{\varphi 0})\varphi'^{1}_{0}{}^{2} \qquad (4.48)$$

The net effect of $K^{\varphi\chi}$ is a change in the 33 component of the equation. The 00 component of the equation is unchanged.

4.4.2 The $h^{2\mu\nu}$ Variation Dynamic Equation

In eq. 4.46 above we add the variation resulting term $h^{2\varphi\chi}\varphi^{2}{}_{s}{}^{\alpha}\varphi^{2}{}_{s\alpha}$ to eq. 5.32 of Blaha (2026d):

$$0 = \{\tfrac{1}{2}(A-1)M^6\Box\} \cdot \\ \cdot\{-(1-A)\, M^6[2(\lambda_{C1} - AM^8)/(1-A)M^6)]h^{1\varphi\chi} + 2(\lambda_{C1} - AM^8 + \lambda_{C3})\eta^{\mu\nu}/((1-A)M^6)\}/(2\lambda_{C3}) + \\ + (\lambda_{C1} - AM^8)\{-(1-A)M^6[\Box + 2(\lambda_{C1} - AM^8)/(1-A)M^6)]h^{1\varphi\chi} + 2(\lambda_{C1} - AM^8 + \\ + \lambda_{C3})\eta^{\mu\nu}/((1-A)M^6)\}/(2\lambda_{C3}) - M^6\Box h^{1\varphi\chi} + (\lambda_{C2} - 2M^8)h^{1\varphi\chi} + (\lambda_{C2} - 2M^8)\eta^{\varphi\chi} - \\ - M^4(g_G^2/4\pi)J^{\varphi\chi} + K^{\varphi\chi} + h^{2\varphi\chi}\varphi^{2}{}_{s}{}^{\alpha}\varphi^{2}{}_{s\alpha} \qquad (4.49)$$

which becomes

$$0 = \{\tfrac{1}{2}(A-1)M^6\Box\} \cdot \\ \cdot\{-(1-A)\, M^6[2(\lambda_{C1} - AM^8)/(1-A)M^6)]h^{1\varphi\chi} + 2(\lambda_{C1} - AM^8 + \lambda_{C3})\eta^{\mu\nu}/((1-A)M^6)]\}/(2\lambda_{C3}) + \\ + (\lambda_{C1} - AM^8)\{-(1-A)M^6[\Box + 2(\lambda_{C1} - AM^8)/(1-A)M^6)]h^{1\varphi\chi} + 2(\lambda_{C1} - AM^8 + \\ + \lambda_{C3})\eta^{\mu\nu}/((1-A)M^6)\}/(2\lambda_{C3}) - M^6\Box h^{1\varphi\chi} + (\lambda_{C2} - 2M^8)h^{1\varphi\chi} + (\lambda_{C2} - 2M^8)\eta^{\varphi\chi} - \\ - M^4(g_G^2/4\pi)J^{\varphi\chi} + K^{\varphi\chi} \qquad (4.50)$$

to leading order after discarding the higher order derivative term.

We can obtain a quadratic equation for $h^{1\varphi\chi}$ by finding the coefficient of the $\Box$ term and the coefficient of the constant factor multiple of $h^{1\varphi\chi}$.

<u>Coefficient of the $\Box h^{1\varphi\chi}$ term</u>

$$\tfrac{1}{2}(A-1)M^6\{-(1-A)M^6 2(\lambda_{C1} - AM^8)/(1-A)M^6)\} + (\lambda_{C1} - AM^8)\{-(1-A)\, M^6\}/(2\lambda_{C3}) - M^6 \qquad (4.51)$$

If A = 0, which we believe is the correct choice, then the coefficient is

$$[\,(\lambda_{C1} - \lambda_{C1} - 1)\,/(2\lambda_{C3})]M^6 = -M^6/(2\lambda_{C3}) \qquad (4.52)$$

<u>Coefficient of the $h^{1\varphi\chi}$ mass term</u>

$$(2(\lambda_{C1} - AM^8)^2)/(2\lambda_{C3}) + \lambda_{C2} - 2M^8 - b_1 b_0^2\delta_{\varphi 3}\delta_{\chi 3} \qquad (4.53)$$

If A = 0 then the coefficient is

$$-\lambda_{C1}{}^2/\lambda_{C3} + \lambda_{C2} - 2M^8 - b_1 b_0{}^2\delta_{\varphi 3}\delta_{\chi 3} \tag{4.54}$$

Then eq. 4.50 above becomes

$$[-M^6/(2\lambda_{C3})) \ \Box + (-\lambda_{C1}{}^2/\lambda_{C3} + \lambda_{C2} - 2M^8 - b_1 b_0{}^2\delta_{\varphi 3}\delta_{\chi 3})]\ h^{1\varphi\chi} \ - M^4(g_G{}^2/4\pi)J^{\varphi\chi} + \text{constants} = 0 \tag{4.55}$$

and the mass corresponding to the time-like h^{100} is

$$m_{10}{}^2 = 2\lambda_{C3}(2M^8 + \lambda_{C1}{}^2/\lambda_{C3} - \lambda_{C2})/M^6 \tag{4.56}$$

and

$$\Box + m_{10}{}^2)h^{1\varphi\chi} + GJ^{\varphi\chi} = 0 \tag{4.57}$$

with

$$G = 2\lambda_{C3}M^{-2}\,(g_G{}^2/4\pi) \tag{4.58}$$

by eq. 4.50. Note the M^{-2} dependence of the effective coupling. Setting

$$2\lambda_{C3}(2M^8 + \lambda_{C1}{}^2/\lambda_{C3} - \lambda_{C2})/M^6 = 0 \tag{4.59}$$

gives a zero mass time-like graviton:

$$m_{10}{}^2 = 0 \tag{4.60}$$

The 33 component in eq. 4.50 increases the mass of the spatial graviton. Taking account of the □'s coefficient in eq. 4.50:

$$-M^6/(2\lambda_{C3}) \tag{4.61}$$

:we find

$$m_{1\text{spatial}}{}^2 = 2\lambda_{C3}(2M^8 + \lambda_{C1}{}^2/\lambda_{C3} - \lambda_{C2})/M^6 + b_1 b_0{}^2(2\lambda_{C3})/\,M^6 \tag{4.62}$$
$$= 2\lambda_{C3}(2M^8 + \lambda_{C1}{}^2/\lambda_{C3} - \lambda_{C2} + b_1 b_0{}^2)/M^6$$

Then eq. 4.50 would imply

$$m_{1\text{spatial}}{}^2 = 2\lambda_{C3} b_1 b_0{}^2/M^6 \tag{4.63}$$
$$m_{1\text{spatial}}{}^2 = 2\lambda_{C3} b_1 b_0{}^2/M^6 \tag{4.64}$$

4.5 B and $h^{1\mu\nu}$ Masses

As a result, for $h^{1\mu\nu}$ we find the masses

$$m_{10}{}^2 = 0 \tag{4.65}$$
$$m_{1\text{spatial}}{}^2 = 2\lambda_{C3}\, b_1 b_0{}^2/M^6 \tag{4.66}$$

and for the $B^{i\mu}$ field we find the masses:

For B^{10} the mass:

$$M_{\varepsilon B0} \ = \ 4b_1 \tag{4.67}$$

For the spatial $\varepsilon_B{}^{0i}$ field mass for coordinates i = 1, 2, 3:

$$M_{\varepsilon Bi} = 16\lambda_B b_0{}^2 + 4b_1 \quad (4.68)$$

The "spatial" h^{1ij} (all i, j with i = j = 0 excluded) and B^{1ij} field masses both depend on the B condensate.

Note b_1 and λ_B both have the dimensions of [mass^2] and λ_{C3} has the dimensions of [mass^6]. Also b_0 is dimensionless.

4.6 Gravitation Constant G From eq. 5.35 we see

The Gravitation constant has inverse mass squared dimensions by eq. 4.50:

$$G = 2\lambda_{C3}M^{-2}(g_G{}^2/4\pi) \quad (4.69)$$

confirming the expected dimension dependence of G.

Note: The $h^{2\varphi\chi}$ variation dynamic equations are unchanged to leading order.

Note: The masses are unchanged from those in Blaha (2026d) when using Higgs vector bosons to leading order. They do affect the dynamics in higher order.

The B condensate induces the breakdown of Lorentz covariance and creates an Absolute Reference Frame – the "rest frame" of the B condensate.

4.7 The Field Equations and Gravitational Radiation Fields

The gravitational dynamic equations that we have found to lowest order have interesting properties of motion. With a proper choice of Cosmological constants the h^{100} has the mass

$$m_{10}{}^2 = 0 \quad (4.65)$$

as expected. Thus the h^{100} field propagates at the speed of light.

For the same choice of Cosmological constants we find all other spatial metric tensor fields have a large mass:

$$m_{1spatial}{}^2 = 2\lambda_{C3}\, b_1 b_0{}^2/M^6 \quad (4.66)$$

for "spatial" h^{1ij} (all i, j with i = j = 0 excluded). Thus these fields propagate at sublight speeds.

<u>Comparison to Conventional Field Equation Derivation</u>

The difference between these cases of time-like and spatial gravitons appears to impact on the conventional derivation of the General Relativistic Field Equations. The conventional derivation is based on Newtonian gravity potential φ and sets the time-time component of the metric tensor:

$$g^{00} \approx 1 + 2\varphi) \tag{4.70}$$

This assumption leads to the further assumption

$$H^{\rho\sigma} = -8\pi G T^{\rho\sigma} \tag{4.71}$$

where $H^{\rho\sigma}$ is an expression in the metric tensor fields. This *assumption* is generalized to the complete set of metric indices of the form of the field equations:

$$R^{\mu\nu} = -8\pi G T^{\mu\nu} \tag{4.72}$$

where $T^{\mu\nu}$ is the energy-momentum tensor.

Thus the field equations of eq. 4.71 are the result of seemingly reasonable assumptions. They have the advantage of a simple Lagrangian density: $\mathscr{L} = g^{\mu\nu}R_{\mu\nu}$.

Our Lagrangian (eq. 4.4) for Enhanced Extended General Relativity (EEGR) is an alternate possibility. *Ab initio* it has very desirable features such as the form of G and the existence of an Absolute Reference Frame. It embodies the Newtonian potential seen in eqs. 4.50 and 4.57. It results in a finite Perturbation Theory.

We conclude that EEGR is a very viable alternative to the conventional formulation of the General Relativity field equations

4.8 Role of Four Condensates

There are four potentials in the preceding derivation:

	Purpose	
$V(\varphi^1) = \sum_{a=1}^{4} \lambda_{\varphi a}(\varphi^1{}_a{}^*\varphi^1{}_a - \varphi^1{}_{0a}{}^2)^2$	set the ElectroWeak Higgs field minimum	(4.8)
$V(\varphi^1{}_s) = \sum_{a=1}^{16} \lambda_{\varphi sa}(\varphi^1{}_{sa}{}^*\varphi^1{}_{sa} - \varphi^1{}_{s0a}{}^2)^2$	set the SU(4) Higgs field minimum	(4.9)
$V_B(B^{1\mu\nu}) = \lambda_B(B^{1\mu}B^1{}_\mu - B_0{}^\mu B_{0\mu})^2$	set the B field minimum	(4.10)
$V_g(g^{1\mu\nu}) = \lambda_R(g^{1\mu\nu}g^1{}_{\mu\nu} - \eta^{\mu\nu}\eta_{\mu\nu})^2$	set the metric tensor minimum	(4.11)

The first two potentials correspond to two potentials that appear in the EWS Lagrangian. The EWS Lagrangian can be directly extended to include the other two potentials with little impact on the EWS dynamic equations in low order.

Each potential has a corresponding condensate. Each condensate has no electric charge and is thus neutral – not directly detectable.

Thus the analogy between our new Enhanced Extended Theory of Gravity (EEGR) and EWS is supported. The net effect of this analogy is to buttress the parts of EEGR in Lagrangian eq. 4.4, which might otherwise be taken as arbitrary.

4.9 Conclusions

There are a number of conclusions that follow from the above calculation:

1. Some of the constants in the Blaha (2026d) calculation were replaced with Higgs Mechanism determined values. They did not change the metric tensor masses in the above calculation. The above calculation attempted to create a Gravitation that was the equivalent of EWS. It is possible to modify terms in the above calculation such that the metric tensor masses would be modified by the Higgs fields $\varphi^{i\mu}$ and $\varphi'^{i}_{s}{}^{\mu}$.

2. The above calculation was to lowest degree in fields so as to have tractable dynamic equations. There were no effects from the Higgs fields $\varphi^{i\mu}$ and $\varphi'^{i}_{s}{}^{\mu}$ to this degree. If higher degree terms were included then the Higgs fields would affect the results.

3. The analogy between our new Enhanced Extended Theory of Gravity (EEGR) and EWS is supported. The parts of EEGR in Lagrangian eq. 4.4, which might be viewed as arbitrary, can be viewed as acceptable due to its features and due to the EWS analogy. Nature often repeats successful features.

4. EEGR has exciting new features in addition to the close analogy with the EWS theory which was the source of its form. Some new features are: support for the Theory of Time which accounts for the universal increase of time, an Absolute reference frame, and the inverse mass squared form of G.

5. An especially important feature is the appearance of quartic terms in the field equations. These terms result in a finite Gravitation Perturbation Theory based on a negative superficial degree of divergence for all EEGR perturbation theory diagrams. Negativity implies a convergent – non-infinite value for a diagram.

EEGR's field equations have their source in EWS. The conventional field equations have their source in the simplicity of the Lagrangian density: $\mathcal{L} = g^{\mu\nu}R_{\mu\nu}$. The reader must make his/her own choice: simplicity (where the EEGR features are not present) or EEGR with its many important features.

5. Theory of Time-Space

We have developed[13] a Theory of Time-Space *in a universe* based on the existence of a B condensate and the above EEGR Theory. In this chapter we extend the theory to the Cosmos outside of universes and develop a quantum formulation of the theory.

5.1 Why Worry About Time?

When we consider time and space we usually take them as given: determined by clocks and yard sticks. Yet we talk of many possibilities in time: slow times, fast times, time reversal to name a few. Time is more than a wall clock. It has important dynamical aspects. It raises the possibility of a dynamic time that is determined by forces in Nature that we have not understood (until now.)

Dynamical theories in Physics often have solutions that specify their time development. However the solutions simply specify what "would happen" as time increases. The continual increase of time is simply assumed to happen in experimental situations.

Why does time increase? It is rightly not part of any Physics solution. It is an extraneous fact tied up with the nature of time and space. Dynamics solutions take it for granted.

An inquiry into the features of space and time, such as this book makes, leads us to find a dynamical mechanism for time (and space) to answer questions such as

1. Why time "always" increases? Why space coordinates do not?
2. Can time go "backwards?"
3. Can time loop?
4. Does time go quickly at times and slowly at other times?

We now consider these aspects of time and space. We suggest that time increase is the consequence of force exerted by the B^{μ} field condensate that we considered earlier. Since the condensate is macroscopic and constant, and since it exists throughout space since the Big Bang, we can consider the generation of continuous time increase (and *apparent* space coordinates constancy) within a classical Physics framework.

The dynamics of time that we develop starts with time t in an Absolute reference frame where the B field condensate is stationary and constant throughout space in an Absolute reference frame. (The time in other reference frames is determined by Lorentz transformations or General Relativity transformations as the case may be.)

---

[13] Blaha (2026d).

We will see that time is a result of a dynamics based on the statistical mechanics of the B condensate.

5.2 Time Within Our Universe

We begin by assuming that increasing time is the result of the interactions associated with the B condensate defined by eq. 4.10. The EEGR theory defines a B condensate as a constant density pervading the entire universe and existing as a constant since the origin of the universe. The B condensate has associated forces – that are evident in the B Lagrangian. They cause interactions with the metric tensor fields and related Higgs fields.

Given this situation, which is implicit in the EEGR formulation, we see that the time-space dynamic equation is not bound by the speed of light for the constant B condensate. Fluctuations in the B condensate would be bound by the speed of light.

The B Lagrangian is

$$\mathscr{L}_{EB} = -\tfrac{1}{4}B^{1\mu\nu}B^{1}{}_{\mu\nu} - V_B(B^{1\mu\nu}) - b_2(B^{2\mu}g^{1}{}_{\mu\nu})^{\dagger}\, g^{1\nu\xi}B^{2}{}_{\xi} - \\ - b_1(B^{1\mu}g^{1}{}_{\mu\nu})^{\dagger}g^{1\nu\xi}B^{1}{}_{\xi} + g_B B^{1\mu}J_\mu + CC$$

We assume the B condensate is in the Absolute Reference Frame described in the previous chapter.

The relevant part of the $B^{1\mu}$ field is the increment field $\varepsilon_B{}^{\mu}$ where

$$B^{1\mu} = B_0{}^{\mu} + \varepsilon_B{}^{\mu} \qquad (4.31)$$

This $\varepsilon_B{}^{\nu}$ field interacts with the energy-momentum current of the entire universe through $g_B B^{1\mu}J_\mu$. From eq. 4.40 we see there are two masses: one for $\nu = 0$ – the time-like $\varepsilon_B{}^{0}$ field

$$M_{\varepsilon B0} = 4b_1 \qquad (4.41)$$

and a spatial $\varepsilon_B{}^{0i}$ field mass for coordinates i = 1, 2, 3:

$$M_{\varepsilon Bi} = 16\lambda_B b_0{}^2 + 4b_1 \qquad (4.42)$$

These masses are an analogue of the W gauge fields Higgs breakdown masses if we set $b_1 = 0$.

We now define a *classical* Lagrangian for time t in the reference frame of the B condensate.[14] Since the B condensate and the increase of time are primarily classical in nature we use a classical formulation:

$$L = -\,mc^2(-dx^{\mu}/d\sigma\, dx_{\mu}/d\sigma)^{1/2} + gB_{\mu}\, dx^{\mu}/d\sigma \qquad (5.1)$$

[14] The B condensate pervades the entire universe. It is usually totally constant with no space-time dependence. Later we consider the consequences of localized "ripples" in the condensate.

where σ is a time-like parameter, B_μ is the B field described earlier, and g is a coupling constant.

After introducing a new parameter T that is *initially* treated as a proper time parameter in the Absolute Reference Frame coordinate system, we find the Lagrangian L implies the classical Euler-Lagrange equations:

$$d\,(mc\, dx^\mu/dT + gB^\mu)/dT - g\, dx^\nu/dT\, \partial^\mu B_\nu \tag{5.2}$$

The non-relativistic equivalent is (Newton's Law)

$$m\, d^2x^\mu/dT^2 = gB^\mu \tag{5.3}$$

where m is a mass-like constant. We use it below to analyze the increase of time (and space?) in general. The variable x^μ/consists of the 4-vector components (t, **x**).

5.2.1 The Continuous, Uniform Increase of Time

We will initially assume the 0th component of the $B^{1\mu}$ field, namely the $B_0 + \varepsilon_B{}^0$ terms, which we will denote as B_0 for compactness, of the B condensate, is constant throughout space in the universe. Eq. 5.3 then implies

$$m\, d^2t/dT^2 = gB_0 \tag{5.4}$$

and has a solution

$$t = \tfrac{1}{2}gT^2B_0/m$$

with velocity

$$v_t = dt/dT = gB_0T/m \tag{5.5}$$

The T dependence of the velocity is at first disconcerting. We re-express the equation as

$$v_t = dt/d(\ln T) = gB_0/m \tag{5.6}$$

using the natural logarithm. The velocity is now constant with respect to ln T.

Remembering Boltzmann's Law for a Physical system:

$$S = k \ln W \tag{5.7}$$

where k is Boltzmann's constant, S is the entropy and W is Boltzmann's multiplicity (the number of real microstates corresponding to the condensate's macrostate in this case) we specify

$$V = v_t/k = dt/d(k\ln T) = gB_0/(mk) \tag{5.8}$$

The velocity of time is now

$$V = dt/dS = gB_0/(mk) \tag{5.9}$$

where

$$S = k\ln T \tag{5.9a}$$

The velocity V is constant giving a uniform growth of time. Eq. 5.9 implies

$$t = gB_0/(mk)\ S \qquad (5.9b)$$

consistent with Statistical Mechanics.

The B condensate entropy S is constant for all the universe's space. S = –∞ at the beginning of the universe and grows – leading to increasing time. We can set

$$S = mkt/(gB_0) + \text{const} \qquad (5.9c)$$

Since the B condensate occupies all the space of the universe, S and t can increase without limit. We now view T as Boltzmann's multiplicity for the B condensate (the number of real microstates corresponding to the condensate's macrostate in this case.)

We thus interpret S as the entropy of the B condensate and define the velocity of time t in terms of it (eq. 5.9)

The B Condensate

The B condensate exists at the creation of the universe (perhaps at the Big Bang point). The condensate consists of a constant distribution of B^1 particles throughout the universe. These particles can self-interact and multiply. Thus the condensate can be viewed as evolving with ever increasing numbers of zero mass B^{10} particles. As a result the number of Boltzmann's multiplicity microstates grows and the entropy grows correspondingly. *The growth in entropy is then without limit. Entropy and thus time in the universe can be defined to be zero at the beginning of the universe and then grow without limit.*

For a given condensate entropy S we find a uniform increase in time t for the ***entire*** *universe in the Absolute coordinate reference frame.*

5.2.2 Constancy of Spatial Coordinates?

We now turn to the question of the constancy of the spatial coordinates in the Absolute reference frame. From eq. 5.3 we have a spatial coordinate x^i satisfying

$$m\ d^2x^i/dT^2 = gB^i \qquad (5.10)$$

with a similar coordinate velocity which is apparently

$$v_i = dx^i d(\ln T) = gB^i/m \qquad (5.11)$$

While the time-like part of B^μ propagates as a zero mass boson by eq 5.19 if $b_l = 0$, we find $B^{1\mu}$'s spatial parts have a massive propagator due to the B condensate according to eq. 4.42. Choosing $b_l = 0$ we have the spatial B^{1i} mass dependent on the condensate:[15]

[15] Note b_l is not equal to B_0.

$$m_{Bspatial}{}^2 = 16\lambda_B b_0{}^2 \tag{5.12}$$

As a result the spatial coordinates' force is diminished. Due to the effective propagator value of $1/m_{Bspatial}{}^2$ eq. 5.11 becomes

$$v_i = dx^i d(\ln T) = gB^i/(mm_{Bspatial}{}^2) \tag{5.13}$$

Following steps similar to those that lead to eq. 5.9 we find

$$\begin{aligned} V_{spatial} = dx^i/dS &= gB^i/(mm_{Bspatial}{}^2) \\ &= gB^i/(16\lambda_B b_0{}^2) \end{aligned} \tag{5.14}$$

Since the B condensate is presumably very massive, the value of $V_{spatial}$ is extremely small. As a result spatial coordinates do not change. This miniscule effect is presumably drowned by quantum smearing.

We conclude time t increases while spatial coordinates do not except possibly for quantum fluctuations. We discuss Quantum Time-Space Theory later. We now address various questions of time.

5.2.3 Can Time Flow Backwards?

Time may flow backward in very special circumstances. First an extremely large mass in a localized spatial region must generate very large B fields using eqs. 5.9 and 5.16 to flip the B field force in the time equation in eq. 5.6 causing the time velocity to go negative. The result would be a velocity to the "past" – time reversal in local regions at best.

5.2.4 Are Time Loops Possible?

Supposing the B fields force can be made to have the familiar form of the simple harmonic oscillator then the time velocity and time itself can display a form of harmonic oscillator – time loops. For example, consider the harmonic force

$$m\, d^2t/dT^2 = -gB_0T \tag{5.15}$$

As a result the time t will display a harmonic behavior of the form:

$$t = \text{const}\sin(\omega T + \varphi) \tag{5.16}$$

for frequency ω and phase φ.

This possibility could only be realized in the neighborhood of a great mass that generates a very large gravity that feeds the growth and variation of the time force.

5.2.5 Variation in Time: Fast and Slow

If the B fields have ripples or fissures (hills and valleys) in the B condensate, then time will slow at low points and increase at high points. Note wall clocks of any sort will respond accordingly with no apparent changes in their clock display. A silent fast and slow in t for everything in the spatial region of the ripples (or fissures?).

5.2.6 Jumps in Time

One can consider condensate force spikes in spatial regions of the B condensate that will cause steps in time – time jumps. (Delta function spikes.) For example consider the force equation with a spike at T_0:

$$m\, d^2t/dT^2 = gB_0\delta(T - T_0) \tag{5.17}$$

As a result the time t will display a step behavior:

$$t = gB_0\, T\, \theta(T - T_0) \tag{5.18}$$

This possibility also requires an extremely large mass concentrated in a spatial region.

5.2.7 Frozen Time

This possibility seems excluded based on the above considerations unless the condensate force is zero in a certain region.

5.2.8 Explosive Growth in Time

Given a large mass generating a very large gravity and a very large B condensate force we can consider a force equation such as

$$m\, d^2t/dT^2 = gB_0 f(T) \tag{5.19}$$

where f(T) is a force that increases with T. As a result t accelerates to larger and larger velocities.

5.2.9 Collapse in Time

If the B field condensate "evaporates" so that the force p(T) decreases then the velocity of time t decreases

$$m\, d^2t/dT^2 = gB_0\, p(T) \tag{5.20}$$

and time comes to a halt eventually. Time depends on the B condensate!

5.2.10 B Condensate Entropy and Time

Macroscopic Entropy is often described as providing an "Arrow of Time" because it never decreases for a Physical system (the Second Law) and often tends to increase. It is based on Boltzmann's Law for a Physical system:

$$S = k \ln W \tag{5.21}$$

where k is Boltzmann's constant, S is the entropy and W is Boltzmann's multiplicity (the number of real microstates corresponding to the condensate's macrostate in this case.)

We now consider the B condensate. It pervades all of space. Since we base the growth of time on it, it must have existed since the beginning of the universe. The constancy of the B condensate gives a similar time evolution to the entire universe with the possible exception of truly massive sources of gravity, and possibly in fissures and ripples in the condensate. (See chapter 12.)

We found the velocity of time above

$$V = dt/dS = gB_0/(mk) \tag{5.9}$$

where we interpreted $S = k \ln T$ as the condensate Entropy of the B condensate. We defined the velocity of time t in terms of it. With this interpretation we can identify

$$T = W_B \tag{5.22}$$

where W_B is the Boltzmann's multiplicity for the B condensate (the number of real microstates corresponding to the B condensate's macrostate in this case) As the number of microstates increases the proper time T increases resulting in an increase in time t.

We therefore define the B condensate Entropy:

$$S_B = k \ln T_B \tag{5.23}$$

The velocity of time is now

$$V = v_t/k = dt/d(k\ln T) = gB_0/(mk) \tag{5.24}$$

or

$$V = dt/dS_B = gB_0/(mk) \tag{5.25}$$

where gB_0 is the B condensate force.

Thus our B condensate has an "Arrow of Time" as well as conventional macrostates have an "Arrow of Time." We therefore have extended the "Arrow of Time" from the macroscopic to the universal microscopic.

S_B sets the ubiquitous, universal "Arrow of Time" and determines time's progress.

5.2.11 Astrophysical Evidence of B Condensate Time

One may ask whether there is any evidence for a B condensate in the universe. Possibly. A truly super massive object could cause the time t to run more quickly in the vicinity of the object. (Eqs. 5.4 and 5.5.) Then chemical reactions in the region could generate light of shorter wavelength. This light would seem of higher frequency to remote detectors. Thus there would be an ultraviolet effect in addition to the normal shifts in light frequency due to the Doppler Effect and Cosmological Expansion. For a depressed condensate region one expects an infrared effect that might be visible in its spectra.

It is possible that B condensate effects can be seen or may have been seen.

5.3 Thoughts on Time

The increase or decrease of time evokes a number of possibilities. One potential possibility that appears and is then discounted is:

Present time cannot be rolled back to Past time. If one attempts to perform such a rollback the past point in time is not reached. Rather the rollback is to a point in time that reflects the events that happened prior to the rollback.

For example consider a sample of uranium at time T, which decays as time progresses, that decreases the uranium content of the sample. If we roll back to time T again, the sample, with its decays, rolls back to time T – but it is not the original sample – rather it is the decayed sample. There is no going back! A thing cannot return to an earlier state.

Appendix 5-A. Condensate Cracks and a Stringy Universe

After the Big Bang, we believe the universe expands. The universe has a uniform B condensate throughout the entire universe according to EEGR theory. It is reasonable to expect the expanding universe to develop fissures in the condensate.

These fissures expand to become cracks. The gravity and B fields within the cracks would necessarily be diminished and crack energy lower than external energy.

In these circumstances it is reasonable to expect that matter and energy would accumulate within the lower energy regions of the cracks, and become stars and galaxies.

Then the cracks would appear to be the Cosmic strings of galaxies that are currently being seen in the universe. Thus the new EEGR theory could account for galaxy strings in the universe.

6. Black Hole Horizon – Time-Space Increase?

There are places in space where very large masses cause the metric tensor to have "unusual" values. One place is the event horizon of a Black Hole. At a point on the event horizon the metric tensor values are:

$$g^{00} = (1 - 2a/r) \rightarrow 0 \tag{6.1}$$

$$g^{ij} = (1 - 2a/r)^{-1} \rightarrow \infty \tag{6.2}$$

as $r \rightarrow 2a$ using the deSitter-Schwarzschild metric.

We now assume that a similar situation holds for the metric tensor solutions of the EEGR theory and examine the consequences for the time change and the space change on the event horizon using Time-Space Theory.

We begin by noting eq. 4.29 implies $B^{2i} \rightarrow 0$ as $g^{ij} \rightarrow \infty$ due to

$$\partial_\mu B^{1\mu\nu} + 4\, b_2 g^{1\mu\nu} g^{1}{}_{\mu}{}^{\xi} B^{2}{}_{\xi} + 2\varphi^{1}{}_{\alpha}\varphi^{1\alpha} B^{2\nu} = 0 \tag{4.29}$$

which contains the term

$$b_2 g^{1\mu\nu} g^{1}{}_{\mu}{}^{\xi} B^{2}{}_{\xi} = b_2 g^{1\nu\xi} B^{2}{}_{\xi} = b_2 g^{1\nu 0} B^{2}{}_{0} + b_2 g^{1\nu i} B^{2}{}_{i} \tag{6.3}$$

The

$$b_2 g^{1\nu i} B^{2}{}_{i} \tag{6.4}$$

term within it rises to infinity unless $B^{2i} \rightarrow 0$. Otherwise eq. 4.29 would not hold.

Having found the limiting case $B^{2i} \rightarrow 0$ we now examine the $\nu = 0$ component of

$$½\, \Box B^{2\nu} + ½\, \Box B^{1\nu} + 16\lambda_B B_0{}^{\nu} B_0{}^{\mu} \varepsilon_\mu + 2\, b_1 g^{1}{}_{\mu}{}^{\nu} g^{1\mu\xi} B^{1}{}_{\xi} + 2\varphi^{2}{}_{\alpha}\varphi^{2\alpha} B^{1\nu} = 0 \tag{6.5}$$

where the increment field $\varepsilon_B{}^{\mu}$ is defined by

$$B^{1\mu} = B_0{}^{\mu} + \varepsilon_B{}^{\mu} \tag{4.31}$$

$$\varphi^{1\nu}(x) = \varphi^{1}{}_{0}{}^{\nu} + \varepsilon^{\nu}(x) \tag{9.5}$$

and

$$\partial_\mu B^{1\mu} = \partial_\mu \varepsilon_B{}^{\mu} = 0 \tag{4.33}$$

$$\partial_\mu B^{2\mu} = 0 \tag{4.34}$$

Eq. 6.5 follows from eqs. 4.30 and 4.32. Noting

$$b_1 g^{1\mu\nu} g^1{}_\mu{}^\xi B^1{}_\xi = b_1 g^{1\nu\xi} B^1{}_\xi = b_1 g^{1\nu 0} B^1{}_0 + b_1 g^{1\nu i} B^1{}_i \tag{6.6}$$

we see the term (for ν = i)

$$b_1 g^{1\nu i} B^1{}_i \tag{6.7}$$

requires

$$B^1{}_i \rightarrow 0 \tag{6.8}$$

Otherwise eq. 6.5 would not hold since the other terms in eq. 6.5 are finite.

Neglecting the quadratic terms in eq. 6.5 and using eqs. 6.4 we find for ν = 0 eq. 6.5 becomes:

$$2\, b_1 g^{1\nu 0} B^1{}_0 + 2\varphi^2{}_\alpha \varphi^{2\alpha} B^{10} = 0 \tag{6.9}$$

which implies

$$B^1{}_0 \rightarrow 0 \tag{6.10}$$

Eqs. 6.8 and 6.10 provide the forces for time increase and spatial increase as we showed in Time-Space Theory using the non-relativistic equivalent of Newton's Law:

$$m\, d^2x^\mu/dT^2 = gB^\mu \tag{5.3}$$

We conclude that time and space do not *change* at points on an Event Horizon.

This approximate behavior of time and space is a zero increase/decrease at the event horizon. It raises several possibilities:

1. Time may decrease within a Black Hole. The Event Horizon then becomes the boundary between increasing and decreasing time.

2. Space may decrease within a Black Hole.

The conclusions obtained here might also apply to "fireWalls."

7. Time-Space Theory Extended to the Cosmos

In chapter 5 we developed Time-Space Theory within one universe. We took that universe to be ours. The theory works for any universe. We now extend the theory to any universe, the Parent universe, and the Cosmos.

We now consider the universe within the Parent universe of Cosmos Theory. We have defined the Cosmos as existing forever without beginning or end. *For the sake of presentation* we have specified the features of the Cosmos as a sequence of events. These events also existed simultaneously for all time and may be viewed as eternal.

The sequence begins with a single mathematical point that proceeds to grow into a fractal string with points along the string representing allowed spaces. The definition of the spaces includes a space-time dimension r and an array of $d = 2^{r+4} = 2^{r'}$ = entries for a HyperCosmos space or $d = 2^{r+3} = 2^{r'}$ = entries for a HyperCosmos Space of the Second Kind where r' depends on the type of the space. Fig. 6.1 displays the sequence of spaces as a DNA-like curve. We view the spaces as fractal in nature.

We define[16] the fractal dimensions of spaces by the size, d, of their dimension arrays. For each space we identify the size as a measure of the number of microstates of the space. Then we can define an entropy:

$$S_{\text{space d}} = k_S \ln_2(d) \tag{7.1}$$

where

$$k_S = 1/\ln(2) \tag{7.2}$$

and ln() is the natural logarithm (to base 2.718) and $\ln_2()$ is the logarithm base 2.

7.1 Cosmos Construction from a Mathematical Point

We can divide the construction of the Cosmos spaces fractal string (See Fig. 7.1 for part of the fractal.) into two parts: the initial construction from a point to a unit line segment, and the second stage construction from a unit line segment through a series of steps that each build a line segment of length $2^{r'}$.

The initial stage uses the identity

$$\sum_{k=-\infty}^{-1} 2^k = 1 \tag{7.3}$$

to construct the unit line segment. The nth step of the second stage uses

$$t_n = \sum_{k=1}^{n} 2^k = 2^n - 1 \tag{7.4}$$

[16] From Blaha (2025f).

to obtain the "running" cumulative length of line segments of the fractal string's n^{th} step:

$$t_n = t_{r'} = \sum_{k=0}^{r'} 2^k = 2^{r'} - 1 \tag{7.4}$$

We can therefore define the entropy of the fractal string at the nth step as

$$S_{space\ n} = k_S \ln_2(t_{r'}) \tag{7.5}$$

The entropy increases with each step. It's values range from 0 at r' = -∞ (n = -∞) at the mathematical point that begins the fractal curve to ∞ for the complete fractal Cosmos curve using eq. 7.4.

7.2 Definition of PseudoTime

Having developed an entropy for the growth steps of the Cosmos Fractal Curve we can now define a kind of time that we will call *PseudoTime*. Using eq. 5.9c as a model we define PseudoTime t as

$$t = k_{ps} S + t_{ps0} \tag{7.6}$$

where k_{ps} and t_{ps0} are constants. The PseudoTime grows with the entropy as expected.

Since the entire Cosmos Fractal Curve with its associated spaces exists instantaneously always without beginning or end, PseudoTime is fictitious in one sense but perhaps useful in tracking the Physical steps of the Cosmos Curve creation while realizing it is instantaneous and eternally existent.

7.3 The Parent Universe

The Cosmos Fractal Curve and its Spaces have always existed. In addition we found it necessary to define a Parent universe that sits at the top of the hierarchy of universes including our own universe. *This universe also exists eternally.* We used its features to specify the Physical bounds on the Cosmos Spaces spectrum.

The Parent universe has the general features of universes such as our own universe. Its distinctive feature is its eternity. In addition it provides the basis for the creation and interactions of child universes as we touched upon in previous books.

The Parent universe has features similar to our universe. Most importantly it must have a General Relativity analogous to our universe's General Relativity as we presented in earlier chapters. In this General Relativity there must be condensates similar to the condensates appearing in chapter 4.

There also must be an equivalent to the B fields described in chapter 4. Thus it makes sense to consider a theory of time in these other universes similar to that of chapter 5. Concomitant with that, a related entropy must exist that is related to time and plays a related role for the increase of time.

We conclude other universes, including the Parent universe, must have similar features to our universe including a General Relativity, B fields, and condensates.

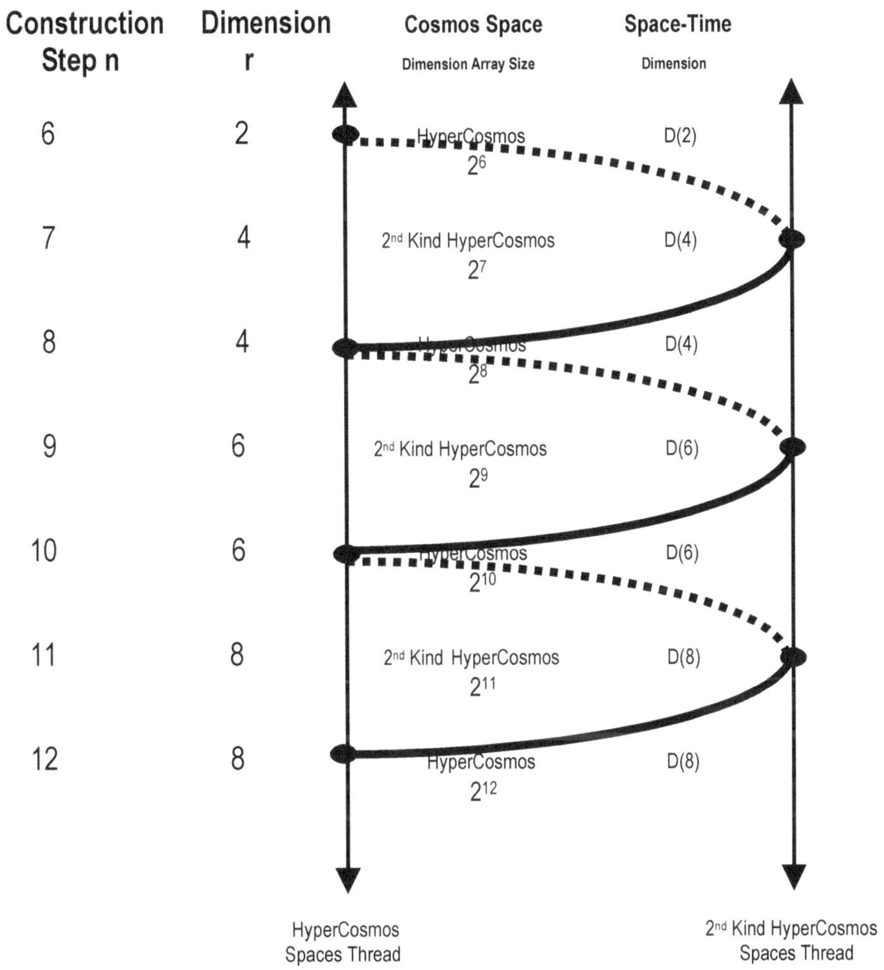

Figure 6.1. Section of the Cosmos Curve showing a double helix form (with the helix mapped to two dimensions.) The two threads of Cosmos spaces and space-times with the "helix" connection between strands are due to the variation in θ. The connecting curved lines form a helix when the 0 spatial dimension of the cutve is considered – an analogue of a double helix similar to DNA. The variation from θ = 0 continuously to θ = π/(2n) takes $D(2^n, \theta)$ from $D(2^{r+4})$ to $D(2^{r+3})$. This change corresponds to a change from n to n + 1 as shown above with solid curving "helix" lines (the "front" view of the helix). The variation from θ = π/(2n) continuously to θ = 0 takes $D(2^n, \theta)$ from $D(2^{r+3})$ to $D(2^{r+4})$. This change corresponds to a change from n to n + 1 as shown above with broken curving "helix" lines (the "rear" view of the helix). The curved lines are not drawn to scale. From Blaha (2025f).

8. EEGR Perturbation Theory

A major problem of attempts to create a Quantum Gravity is the infinities that crop up in Perturbation Theory calculations. The conventional field equations have their source in the simplicity of the Lagrangian density: $\mathscr{L} = g^{\mu\nu}R_{\mu\nu}$ with its quadratic field equations. There are numerous excellent books and monographs on classical gravity and a large literature on quantum gravity. Therefore our discussion will assume the reader is familiar with the conventional theory.

EEGR has quartic field equations shown in chapter 4 in sections 4.4.1 and 4.4.2. These quartic equations generate Feynman propagators with a $1/k^4$ Fourier representation form which we will consider later.

8.1 Conventional Graviton Propagator

In Blaha (2005a) we presented the "transverse" graviton propagator as a time-ordered product of field operators:[17]

$$i\Delta_{F2}(y_1 - y_2)_{\lambda\tau\rho\sigma} = <0|T(h_{\lambda\tau}(y_1), h_{\rho\sigma}(y_2)|0> \qquad (9.46)$$

$$= -i\int \frac{d^4k\, e^{-ik\cdot(y_1 - y_2)}\, b_{\lambda\tau\rho\sigma}(k)}{(2\pi)^4\,(k^2 + i\varepsilon)}$$

where **k** is a spatial 3-vector, and where $b_{\mu\nu\rho\sigma}(k)$ is a function of k only:

$$b_{\alpha\beta\mu\nu}(k) = \tfrac{1}{2}[(\eta_{\alpha\mu} - k_\alpha k_\mu/\mathbf{k}^2)(\eta_{\beta\nu} - k_\beta k_\nu/\mathbf{k}^2) + (\eta_{\alpha\nu} - k_\alpha k_\nu/\mathbf{k}^2)(\eta_{\beta\mu} - k_\beta k_\mu/\mathbf{k}^2) -$$

$$- (\eta_{\alpha\beta} - k_\alpha k_\beta/\mathbf{k}^2)(\eta_{\mu\nu} - k_\mu k_\nu/\mathbf{k}^2)] \qquad (9.47)$$

where $\alpha, \beta, \mu, \nu = 0, 1, 2, 3$.

The conventional quantum gravitational interaction also has an "instantaneous" part (similar to the instantaneous Coulomb interaction of QED) in addition to the transverse interaction embodied in eq. 9.46. This "instantaneous" interaction contains the Newtonian potential as its large distance limit. The sum of the instantaneous interaction and the transverse interaction gives the total gravitational interaction.

[17] References to equations 9. … refer to equations in Blaha (2005a).

The above graviton propagator has the form given in eq. 6.102 of Blaha (2005a). The leading momentum dependence of the graviton propagator in momentum space is

$$i\Delta_{F2}{}^{TT}(p)_{\lambda\tau\rho\sigma} \backsim p^{-6} \tag{9.48}$$

The superficial degree of divergence D for each diagram in Gravitation Perturbation Theory indicates the large momentum behavior of the calculation corresponding to the diagram. For positive D we obtain a divergent result. For negative D the large momentum behavior is finite. For D = 0 the result is not determined.

For pure *conventional* Quantum Gravity DeWitt[18] finds the superficial degree of divergence of a diagram to be:

$$D = -2L_i + 2\sum_n V_n + 4K \tag{9.62}$$

where L_i is the number of internal lines, V_n is the number of n-pronged vertices, and K is the number of independent momentum integrations. DeWitt points out

$$K = L_i - \sum_n V_n + 1 \tag{9.63}$$

Thus the superficial degree of divergence of a *conventional* Quantum Gravity diagram is:

$$D = 2(K + 1) \tag{9.64}$$

for $K \geq 1$, displaying an ever increasing degree of divergence as the order of the diagram increases.

Conventional Gravitation Theory has a divergent Perturbation Theory.

8.2 EEGR Graviton Propagator

The EEGR field equations imply the Feynman propagators have the form

$$i\Delta_{FEEGR}(y_1 - y_2)_{\lambda\tau\rho\sigma} = \langle 0|T(h^i{}_{\lambda\tau}(y_1), h^j{}_{\rho\sigma}(y_2)|0\rangle$$

$$= -i \int \frac{d^4k\, e^{-ik\cdot(y_1 - y_2)}\, b_{\lambda\tau\rho\sigma}(k)}{(2\pi)^4\,(k^4 + Ak^2 - B + i\varepsilon)}$$

where A and B are constants.

---

[18] B. S. DeWitt, Phys. Rev. **162**, 1239 (1967).

The field equations and the propagators have an especially important feature: the appearance of quartic terms in the field equations and k^4 terms in the propagator representation. These terms result in a finite EEGR Gravitation Perturbation Theory based on a negative superficial degree of divergence for all EEGR perturbation theory diagrams. Negativity implies a convergent – non-infinite value for a diagram.

In the case of EEGR perturbation theory the superficial degree of divergence of a diagram is:

$$D_{TT} = -6L_i + 2\sum_n V_n + 4K \qquad (8.1)$$

Since any diagram with a loop has $L_i \geq 1$ and $\sum_n V_n \geq 1$ we see that $D \leq -2$. Thus *all* EEGR diagrams are convergent.

8.3 Two-Tier Quantum Gravity: Finite!

The above view of the EEGR Perturbation Theory implies all perturbation theory diagrams are finite with no infinities. We make the finiteness "stronger" with our Two-Tier Theory. In Blaha (2005a) we presented Two-Tier Perturbation Theory in detail. Here we will consider important points of this perturbation theory.

We developed Two-Tier Theory in a series of books in the 2000 – 2010 decade. It was an original effort to resolve perturbation theory infinities for all diagrams of all Quantum Field Theories. It even resolved the infinities of the three fermion triangle diagrams as I showed Roman Jackiw (MIT) at that time.[19]

We originated Two-Tier Theory around 2000 AD based on a simple mechanical model of a harmonic oscillator residing on a harmonic oscillator. Thus the name Two-Tier – two mechanical levels, which we used as a model to create Quantum Two-Tier Theory.[20]

We will begin by establishing the general form of Two-Tier classical General Relativity and then proceed to define a quantization procedure. We will work in Minkowski space with three space and one time dimension. The flat-space metric $\eta_{\alpha\beta}$ is defined as diagonal with $\eta_{00} = 1$ and $\eta_{ij} = -\delta_{ij}$ for i, j = 1, 2, 3.

8.4 Two-Tier General Relativity

In developing Quantum Gravity we make the same ansatz that we have made throughout our development of Two-Tier quantum field theories: all field expressions are functions of an X coordinate field system, which in turn are functions of the "ordinary" y space-time coordinate system. Two-Tier Theory of Quantum Gravity is invariant under General Relativistic transformations. The dynamical field equations, which are strictly functions of the X coordinates, are covariant under general relativistic

---

[19] Two-Tier Theory and our PseudoQuantum Field Theory, which we developed in the 1970's for r potential Quark confinement and to resolve General Relativity accelerated frame transformation issues with particle interpretation, together give a new character to Quantum Theory as we have shown periodically over the past 26 years.

[20] See Blaha (2002).

transformations. The rationale for these assumptions is described in detail in chapter 10 of Blaha (2005a).

We define all coordinates in a Lagrangian (either EEGR or some other Quantum Field Theory or PseudoQuantum Field Theory) as

$$X^{\mu}(y) = y^{\mu} + i\, Y^{\mu}(y)/M_c^{\,2} \tag{8.2}$$

where $Y^{\mu}(y)$ is a zero mass electromagnetic-like field and M_c is a constant.

The algebraic manipulation of tensor indices is the same as that of EEGR. Details can be found in Blaha (2205a).

The "transverse" graviton propagator can be represented as a time-ordered product of field operators. For a conventional gravity theory with a quadratic field equation:

$$i\Delta_{F2}{}^{TT}(y_1 - y_2)_{\lambda\tau\rho\sigma} = <0|T(h_{\lambda\tau}(X(y_1)), h_{\rho\sigma}(X(y_2)))|0> \tag{8.3}$$

$$= -i \int \frac{d^4k\, e^{-ik\cdot(y_1 - y_2)}\, b_{\lambda\tau\rho\sigma}(k) R(\mathbf{k}, y_1 - y_2)}{(2\pi)^4\, (k^2 + i\varepsilon)}$$

where $R(\mathbf{k}, y_1 - y_2)$ is an exponential damping factor appearing in propagators throughout Two-Tier theories, $\mathbf{k}$ is a spatial 3-vector, and where $b_{\mu\nu\rho\sigma}(k)$ is a function of k only:

$$b_{\alpha\beta\mu\nu}(k) = \tfrac{1}{2}[(\eta_{\alpha\mu} - k_\alpha k_\mu/\mathbf{k}^2)(\eta_{\beta\nu} - k_\beta k_\nu/\mathbf{k}^2) + (\eta_{\alpha\nu} - k_\alpha k_\nu/\mathbf{k}^2)(\eta_{\beta\mu} - k_\beta k_\mu/\mathbf{k}^2) -$$

$$- (\eta_{\alpha\beta} - k_\alpha k_\beta/\mathbf{k}^2)(\eta_{\mu\nu} - k_\mu k_\nu/\mathbf{k}^2)] \tag{8.4}$$

where α, β, μ, ν = 0, 1, 2, 3. The factor $R(\mathbf{k}, y_1 - y_2)$ guarantees finite perturbation theory diagrams.

The quantum gravitational interaction of this theory are made convergent by the application of Two-Tier Quantum Theory.

EEGR Two-Tier Perturbation Theory

For EEGR theory with quartic field equations:

$$i\Delta_{F2}{}^{TT}(y_1 - y_2)_{\lambda\tau\rho\sigma} = <0|T(h_{\lambda\tau}(X(y_1)), h_{\rho\sigma}(X(y_2)))|0> \tag{8.5}$$

$$= -i \int \frac{d^4k\, e^{-ik\cdot(y_1 - y_2)}\, b_{\lambda\tau\rho\sigma}(k) R(\mathbf{k}, y_1 - y_2)}{(2\pi)^4\, (k^4 + Ak^2 - B + i\varepsilon)}$$

where R(**k,** $y_1 - y_2$) is an exponential damping factor, **k** is a spatial 3-vector, and where $b_{\mu\nu\rho\sigma}(k)$ is a function of k. EEGR perturbation theory is finite in itself.

EEGR perturbation theory can be directly extended to be a Two-Tier Quantum Theory. Then one might say EEGR perturbation theory is "doubly" finite.

8.5 Two-Tier Unitarity

A possible source for concern might be unitarity. Some factors that eliminate this possibility are:

1. Two-Tier Theories agree with "normal" theories at energies below M_c.

2. Unitarity has never been verified. Ultra high energy phenomena are not experimentally accessible.

3. If there were a violation of unitarity in the usual formulation of unitarity, one can simply renormalize it to make probabilities sum to unity. Since low energy results for Two-Tier calculations are the same as the conventional results, deviations due to such a renormalization of probabilities, should that prove necessary, can only affect very high energy calculations.

Therefore unitarity is not an issue. In any case EEGR has a finite perturbation theory.

REFERENCES

Achiever, N. I., Frank, A. H. (tr), 1962, *The Calculus of Variations* (Blaisdell Publishing, New York, 1962).
Bjorken, J. D., Drell, S. D., 1964, *Relativistic Quantum Mechanics* (McGraw-Hill, New York, 1965).
Bjorken, J. D., Drell, S. D., 1965, *Relativistic Quantum Fields* (McGraw-Hill, New York, 1965).
Blaha, S., 1995, *C++ for Professional Programming* (International Thomson Publishing, Boston, 1995).
_______, 1998, *Cosmos and Consciousness* (Pingree-Hill Publishing, Auburn, NH, 1998 and 2002).
_______, 2002, *A Finite Unified Quantum Field Theory of the Elementary Particle Standard Model and Quantum Gravity Based on New Quantum Dimensions™ & a New Paradigm in the Calculus of Variations* (Pingree-Hill Publishing, Auburn, NH, 2002).
_______, 2004, *Quantum Big Bang Cosmology: Complex Space-time General Relativity, Quantum Coordinates™ Dodecahedral Universe, Inflation, and New Spin 0, ½, 1 & 2 Tachyons & Imagyons* (Pingree-Hill Publishing, Auburn, NH, 2004).
______, *2005a, Quantum Theory of the Third Kind: A New Type of Divergence-free Quantum Field Theory Supporting a Unified Standard Model of Elementary Particles and Quantum Gravity based on a New Method in the Calculus of Variations* (Pingree-Hill Publishing, Auburn, NH, 2005).
______, 2005b, *The Metatheory of Physics Theories, and the Theory of Everything as a Quantum Computer Language* (Pingree-Hill Publishing, Auburn, NH, 2005).
______, 2005c, *The Equivalence of Elementary Particle Theories and Computer Languages: Quantum Computers, Turing Machines, Standard Model, Superstring Theory, and a Proof that Gödel's Theorem Implies Nature Must Be Quantum* (Pingree-Hill Publishing, Auburn, NH, 2005).
______, 2006a, *The Foundation of the Forces of Nature* (Pingree-Hill Publishing, Auburn, NH, 2006).
______, 2006b, *A Derivation of ElectroWeak Theory based on an Extension of Special Relativity; Black Hole Tachyons; & Tachyons of Any Spin.* (Pingree-Hill Publishing, Auburn, NH, 2006).
_______, 2007a, *Physics Beyond the Light Barrier: The Source of Parity Violation, Tachyons, and A Derivation of Standard Model Features* (Pingree-Hill Publishing, Auburn, NH, 2007).
______, 2007b, *The Origin of the Standard Model: The Genesis of Four Quark and Lepton Species, Parity Violation, the ElectroWeak Sector, Color SU(3), Three Visible Generations of Fermions, and One Generation of Dark Matter with Dark Energy* (Pingree-Hill Publishing, Auburn, NH, 2007).
______, *2008a, A Direct Derivation of the Form of the Standard Model From GL(16) (Pingree-Hill Publishing, Auburn, NH, 2008).*
______, 2008b, *A Complete Derivation of the Form of the Standard Model With a New Method to Generate Particle Masses Second Edition* (Pingree-Hill Publishing, Auburn, NH, 2008)
______, 2009, *The Algebra of Thought & Reality: The Mathematical Basis for Plato's Theory of Ideas, and Reality Extended to Include A Priori Observers and Space-Time Second Edition* (Pingree-Hill Publishing, Auburn, NH, 2009).
______, 2009a, *Bright Stars, Bright Universe* (Pingree-Hill Publishing, Auburn, NH, 2008)
______, 2010a, *Operator Metaphysics: A New Metaphysics Based on a New Operator Logic and a New Quantum Operator Logic that Lead to a Mathematical Basis for Plato's Theory of Ideas and Reality* (Pingree-Hill Publishing, Auburn, NH, 2010).
______, 2010b, *The Standard Model's Form Derived from Operator Logic, Superluminal Transformations and GL(16)* (Pingree-Hill Publishing, Auburn, NH, 2010).
______, 2010c, *SuperCivilizations: Civilizations as Superorganisms* (McMann-Fisher Publishing, Auburn, NH, 2010).
______, 2011a, *21st Century Natural Philosophy Of Ultimate Physical Reality* (McMann-Fisher Publishing, Auburn, NH, 2011).
______, 2011b, *All the Universe! Faster Than Light Tachyon Quark Starships & Particle Accelerators with the LHC as a Prototype Starship Drive Scientific Edition* (Pingree-Hill Publishing, Auburn, NH, 2011).
______, 2011c, *From Asynchronous Logic to The Standard Model to Superflight to the Stars* (Blaha Research, Auburn, NH, 2011).

______, 2012a, *From Asynchronous Logic to The Standard Model to Superflight to the Stars volume 2: Superluminal CP and CPT, U(4) Complex General Relativity and The Standard Model, Complex Vierbein General Relativity, Kinetic Theory, Thermodynamics* (Blaha Research, Auburn, NH, 2012).
______, 2012b, *Standard Model Symmetries, And Four And Sixteen Dimension Complex Relativity; The Origin Of Higgs Mass Terms* (Blaha Reasearch, Auburn, NH, 2012).
______, 2013a, *Multi-Stage Space Guns, Micro-Pulse Nuclear Rockets, and Faster-Than-Light Quark-Gluon Ion Drive Starships* (Blaha Research, Auburn, NH, 2013).
______, 2013b, *The Bridge to Dark Matter; A New Sibling Universe; Dark Energy; Inflatons; Quantum Big Bang; Superluminal Physics; An Extended Standard Model Based on Geometry* (Blaha Reasearch, Auburn, NH, 2013).
______, 2014a, *Universes and Megaverses: From a New Standard Model to a Physical Megaverse; The Big Bang; Our Sibling Universe's Wormhole; Origin of the Cosmological Constant, Spatial Asymmetry of the Universe, and its Web of Galaxies; A Baryonic Field between Universes and Particles; Megaverse Extended Wheeler-DeWitt Equation* (Blaha Reasearch, Auburn, NH, 2014).
______, 2014b, *All the Megaverse! Starships Exploring the Endless Universes of the Cosmos Using the Baryonic Force* (Blaha Research, Auburn, NH, 2014).
______, 2014c, *All the Megaverse! II Between Megaverse Universes: Quantum Entanglement Explained by the Megaverse Coherent Baryonic Radiation Devices – PHASERs Neutron Star Megaverse Slingshot Dynamics Spiritual and UFO Events, and the Megaverse Microscopic Entry into the Megaverse* (Blaha Research, Auburn, NH, 2014).
______, 2015a, *PHYSICS IS LOGIC PAINTED ON THE VOID: Origin of Bare Masses and The Standard Model in Logic, U(4) Origin of the Generations, Normal and Dark Baryonic Forces, Dark Matter, Dark Energy, The Big Bang, Complex General Relativity, A Megaverse of Universe Particles* (Blaha Research, Auburn, NH, 2015).
______, 2015b, *PHYSICS IS LOGIC Part II: The Theory of Everything, The Megaverse Theory of Everything, U(4)⊗U(4) Grand Unified Theory (GUT), Inertial Mass = Gravitational Mass, Unified Extended Standard Model and a New Complex General Relativity with Higgs Particles, Generation Group Higgs Particles* (Blaha Research, Auburn, NH, 2015).
____, 2015c, *The Origin of Higgs ("God") Particles and the Higgs Mechanism: Physics is Logic III, Beyond Higgs – A Revamped Theory With a Local Arrow of Time, The Theory of Everything Enhanced, Why Inertial Frames are Special, Universes of the Mind* (Blaha Research, Auburn, NH, 2015).
______, 2015d, *The Origin of the Eight Coupling Constants of The Theory of Everything: U(8) Grand Unified Theory of Everything (GUTE), S^8 Coupling Constant Symmetry, Space-Time Dependent Coupling Constants, Big Bang Vacuum Coupling Constants, Physics is Logic IV* (Blaha Research, Auburn, NH, 2015).
______, 2016a, *New Types of Dark Matter, Big Bang Equipartition, and A New U(4) Symmetry in the Theory of Everything: Equipartition Principle for Fermions, Matter is 83.33% Dark, Penetrating the Veil of the Big Bang, Explicit QFT Quark Confinement and Charmonium, Physics is Logic V* (Blaha Research, Auburn, NH, 2016).
______, 2016b, *The Periodic Table of the 192 Quarks and Leptons in The Theory of Everything: The U(4) Layer Group, Physics is Logic VI* (Blaha Research, Auburn, NH, 2016).
______, 2016c, *New Boson Quantum Field Theory, Dark Matter Dynamics, Dark Matter Fermion Layer Mixing, Genesis of Higgs Particles, New Layer Higgs Masses, Higgs Coupling Constants, Non-Abelian Higgs Gauge Fields, Physics is Logic VII* (Blaha Research, Auburn, NH, 2016).
______, 2016d, *Unification of the Strong Interactions and Gravitation: Quark Confinement Linked to Modified Short-Distance Gravity; Physics is Logic VIII* (Blaha Research, Auburn, NH, 2016).
______, 2016e, *MoND: Unification of the Strong Interactions and Gravitation II, Quark Confinement Linked to Large-Scale Gravity, Physics is Logic IX* (Blaha Research, Auburn, NH, 2016).
______, 2016f, *CQ Mechanics: A Unification of Quantum & Classical Mechanics, Quantum/Semi-Classical Entanglement, Quantum/Classical Path Integrals, Quantum/Classical Chaos* (Blaha Research, Auburn, NH, 2016).
______, 2016g, *GEMS Unified Gravity, ElectroMagnetic and Strong Interactions: Manifest Quark Confinement, A Solution for the Proton Spin Puzzle, Modified Gravity on the Galactic Scale* (Pingree Hill Publishing, Auburn, NH, 2016).
______, 2016h, *Unification of the Seven Boson Interactions based on the Riemann-Christoffel Curvature Tensor* (Pingree Hill Publishing, Auburn, NH, 2016).
______, 2017a, *Unification of the Eleven Boson Interactions based on 'Rotations of Interactions'* (Pingree Hill Publishing, Auburn, NH, 2017).
______, 2017b, *The Origin of Fermions and Bosons, and Their Unification* (Pingree Hill Publishing, Auburn, NH, 2017).
______, 2017c, *Megaverse: The Universe of Universes* (Pingree Hill Publishing, Auburn, NH, 2017).
______, 2017d, *SuperSymmetry and the Unified SuperStandard Model* (Pingree Hill Publishing, Auburn, NH, 2017).

______, 2017e, *From Qubits to the Unified SuperStandard Model with Embedded SuperStrings: A Derivation* (Pingree Hill Publishing, Auburn, NH, 2017).
______, 2017f, *The Unified SuperStandard Model in Our Universe and the Megaverse: Quarks, ... ,* (Pingree Hill Publishing, Auburn, NH, 2017).
______, 2018a, *The Unified SuperStandard Model and the Megaverse SECOND EDITION A Deeper Theory based on a New Particle Functional Space that Explicates Quantum Entanglement Spookiness (Volume 1)* (Pingree Hill Publishing, Auburn, NH, 2018).
______, 2018b, *Cosmos Creation: The Unified SuperStandard Model, Volume 2, SECOND EDITION* (Pingree Hill Publishing, Auburn, NH, 2018).
______, 2018c, *God Theory (*Pingree Hill Publishing, Auburn, NH, 2018).
______, 2018d, *Immortal Eye: God Theory: Second Edition* (Pingree Hill Publishing, Auburn, NH, 2018).
______, 2018e, *Unification of God Theory and Unified SuperStandard Model THIRD EDITION* (Pingree Hill Publishing, Auburn, NH, 2018).
______, 2019a, *Calculation of: QED α = 1/137, and Other Coupling Constants of the Unified SuperStandard Theory* (Pingree Hill Publishing, Auburn, NH, 2019).
______, 2019b, *Coupling Constants of the Unified SuperStandard Theory SECOND EDITION* (Pingree Hill Publishing, Auburn, NH, 2019).
______, 2019c, *New Hybrid Quantum Big_Bang–Megaverse_Driven Universe with a Finite Big Bang and an Increasing Hubble Constant* (Pingree Hill Publishing, Auburn, NH, 2019).
______, 2019d, *The Universe, The Electron and The Vacuum* (Pingree Hill Publishing, Auburn, NH, 2019).
______, 2019e, *Quantum Big Bang – Quantum Vacuum Universes (Particles)* (Pingree Hill Publishing, Auburn, NH, 2019).
______, 2019f, *The Exact QED Calculation of the Fine Structure Constant Implies ALL 4D Universes have the Same Physics/Life Prospects* (Pingree Hill Publishing, Auburn, NH, 2019).
______, 2019g, *Unified SuperStandard Theory and the SuperUniverse Model: The Foundation of Science* (Pingree Hill Publishing, Auburn, NH, 2019).
______, 2020a, *Quaternion Unified SuperStandard Theory (The QUeST) and Megaverse Octonion SuperStandard Theory (MOST)* (Pingree Hill Publishing, Auburn, NH, 2020).
______, 2020b, *United Universes Quaternion Universe - Octonion Megaverse* (Pingree Hill Publishing, Auburn, NH, 2020).
______, 2020c, *Unified SuperStandard Theories for Quaternion Universes & The Octonion Megaverse* (Pingree Hill Publishing, Auburn, NH, 2020).
______, 2020d, *The Essence of Eternity: Quaternion & Octonion SuperStandard Theories* (Pingree Hill Publishing, Auburn, NH, 2020).
______, 2020e, *The Essence of Eternity II* (Pingree Hill Publishing, Auburn, NH, 2020).
______, 2020f, *A Very Conscious Universe* (Pingree Hill Publishing, Auburn, NH, 2020).
______, 2020g, *Hypercomplex Universe* (Pingree Hill Publishing, Auburn, NH, 2020).
______, 2020h, *Beneath the Quaternion Universe* (Pingree Hill Publishing, Auburn, NH, 2020).
______, 2020i, *Why is the Universe Real? From Quaternion & Octonion to Real Coordinates* (Pingree Hill Publishing, Auburn, NH, 2020).
______, 2020j, *The Origin of Universes: of Quaternion Unified SuperStandard Theory (QUeST); and of the Octonion Megaverse (UTMOST)* (Pingree Hill Publishing, Auburn, NH, 2020).
______, 2020k, *The Seven Spaces of Creation: Octonion Cosmology* (Pingree Hill Publishing, Auburn, NH, 2020).
______, 2020l, *From Octonion Cosmology to the Unified SuperStandard Theory of Particles* (Pingree Hill Publishing, Auburn, NH, 2020).
______, 2021a, *Pioneering the Cosmos* (Pingree Hill Publishing, Auburn, NH, 2021).
______, 2021b, *Pioneering the Cosmos II* (Pingree Hill Publishing, Auburn, NH, 2021).
______, 2021c, *Beyond Octonion Cosmology* (Pingree Hill Publishing, Auburn, NH, 2021).
______, 2021d, *Universes are Particles* (Pingree Hill Publishing, Auburn, NH, 2021).
______, 2021e, *Octonion-like dna-based life, Universe expansion is decay, Emerging New Physics* (Pingree Hill Publishing, Auburn, NH, 2021).
______, 2021f, *The Science of Creation New Quantum Field Theory of Spaces* (Pingree Hill Publishing, Auburn, NH, 2021).
______, 2021g, *Quantum Space Theory With Application to Octonion Cosmology & Possibly To Fermionic Condensed Matter* (Pingree Hill Publishing, Auburn, NH, 2021).
______, 2021h, *21st Century Natural Philosophy of Octonion Cosmology , and Predestination, Fate, and Free Will* (Pingree Hill Publishing, Auburn, NH, 2021).

______, 2021i, *Beyond Octonion Cosmology II : Origin of the Quantum; A New Generalized Field Theory (GiFT); A Proof of the Spectrum of Universes; Atoms in Higher Universes* (Pingree Hill Publishing, Auburn, NH, 2021).
______, 2021j, *Integration of General Relativity and Quantum Theory: Octonion Cosmology, GiFT, Creation/Annihilation Spaces CASe, Reduction of Spaces to a Few Fermions and Symmetries in Fundamental Frames* (Pingree Hill Publishing, Auburn, NH, 2021).
______, 2022a, *New View of Octonion Cosmology Based on the Unification of General Relativity and Quantum Theory* (Pingree Hill Publishing, Auburn, NH, 2022).
______, 2022b, *The Dust Beneath Hypercomplex Cosmology* (Pingree Hill Publishing, Auburn, NH, 2022).
______, 2022c, *Passing Through Nature to Eternity: ProtoCosmos, HyperCosmos, Unified SuperStandard Theory* (Pingree Hill Publishing, Auburn, NH, 2022).
______, 2022d, *HyperCosmos Fractionation and Fundamental Reference Frame Based Unification: Particle Inner Space Basis of Parton and Dual Resonance Models* (Pingree Hill Publishing, Auburn, NH, 2022).
______, 2022e, *A New UniDimension ProtoCosmos and SuperString F-Theory Relation to the HyperCosmos* (Pingree Hill Publishing, Auburn, NH, 2022).
______, 2022f, *The Cosmic Panorama: ProtoCosmos, HyperCosmos,Unified SuperStandard Theory (UST) Derivation* (Pingree Hill Publishing, Auburn, NH, 2022).
_____, 2022g, *Ultimate Origin: ProtoCosmos and HyperCosmos* (Pingree Hill Publishing, Auburn, NH, 2022).
______, 2023a, *UltraUnification and the Generation of the Cosmos* (Pingree Hill Publishing, Auburn, NH, 2023).
______, 2023b, *God and and Cosmos Theory* (Pingree Hill Publishing, Auburn, NH, 2023).
______, 2023c, *A New Completely Geometric SU(8) Cosmos Theory; New PseudoFermion Fields; Fibonacci-like Dimension Arrays; Ramsey Number Approximation* (Pingree Hill Publishing, Auburn, NH, 2023).
______, 2023d, *Newton's Apple is Now the Fermion* (Pingree Hill Publishing, Auburn, NH, 2023).
______, 2023e,*Cosmos Theory: The Sub-Particle Gambol Model* (Pingree Hill Publishing, Auburn, NH, 2023).
_____, 2024a, *Cosmos-Universe-Particle-Gambol Theory* (Pingree Hill Publishing, Auburn, NH, 2024).
_____, 2024b, *Fractal Cosmos Theory* (Pingree Hill Publishing, Auburn, NH, 2024).
_____, 2024c, *Fractal Cosmic Curve: Tensor-Based CosmosTheory* (Pingree Hill Publishing, Auburn, NH, 2024).
_____, 2024d, *The Eternal Form of Cosmos Theory* (Pingree Hill Publishing, Auburn, NH, 2024).
_____, 2024e, *The Eternal Form of Cosmos Theory Third Edition* (Pingree Hill Publishing, Auburn, NH, 2024).
_____, 2024f, *Fundamental Constants of Cosmos Theory and The Standard Model* (Pingree Hill Publishing, Auburn, NH, 2024).
_____, 2024g, *Quark, Lepton, W and Z Masses of Cosmos Theory and The Standard Model* (Pingree Hill Publishing, Auburn, NH, 2024).
_____, 2024h, *Geometric Cosmos Geometric Universe* (Pingree Hill Publishing, Auburn, NH, 2024).
_____, 2024i, *Particles and Universes of Cosmos Theory* (Pingree Hill Publishing, Auburn, NH, 2024).
_____, 2024j, *Unification of the Subluminal and the Superluminal in Cosmos Theory* (Pingree Hill Publishing, Auburn, NH, 2024).
_____, 2024k, *The Dawn of Dynamic Cosmos Dimension Arrays* (Pingree Hill Publishing, Auburn, NH, 2024).
_____, 2024ℓ, *Structure and Dynamics of Cosmos Theory and the Unified SuperStandard Theory* (Pingree Hill Publishing, Auburn, NH, 2024).
_____, 2025a, *Black Holes, White Holes, and Superluminal Starship* (Pingree Hill Publishing, Auburn, NH, 2025).
_____, 2025b, *The Sequence Theory Foundation of Cosmos Theory and Gambol Theory* (Pingree Hill Publishing, Auburn, NH, 2025).
____, 2025c, *BEYOND TIME AND SPACE Fermion Strata & Consistency Conditions for Sequences, Cosmos Spaces and Gambols* (Pingree Hill Publishing, Auburn, NH, 2025).
____, 2025d, *The C o s m o s as a Physics Organism* (Blaha Research, Auburn, NH, 2025).
____, 2025e, *The Root of Reality* (Blaha Research, Auburn, NH, 2025).
____, 2025f, *The Fractal Cosmos* (Blaha Research, Auburn, NH, 2025).
____, 2025g, *The Fractal Cosmos II* (Blaha Research, Auburn, NH, 2025).
____, 2025h, *The Fractal Cosmos III* (Blaha Research, Auburn, NH, 2025).
____, 2025i, *Quantum Coalescent Cosmos Fractal Cosmos IV* (Blaha Research, Auburn, NH, 2025).
____, 2025j, *The Entropic Basis of Cosmos Theory The Agent of Origin* (Blaha Research, Auburn, NH, 2025).
____, 2026a, *Cosmos Proof of the Existence of God* (Blaha Research, Auburn, NH, 2026).
____, 2026b, *The Edifice of Cosmos Theory with United Higgs and Quark Confinement Mechanism and a Lumiverse* (Blaha Research, Auburn, NH, 2026).

____, 2026c, *Cosmos ElectroWeakStrong (EWS) Theory and Lumiverse Superspins* (Blaha Research, Auburn, NH, 2026).
____, 2026d, *New Dynamic Theory of General Relativity and a New Dynamic Theory of Time and Space* (Blaha Research, Auburn, NH, 2026).
Eddington, A. S., 1952, *The Mathematical Theory of Relativity* (Cambridge University Press, Cambridge, U.K., 1952).
Fant, Karl M., 2005, *Logically Determined Design: Clockless System Design With NULL Convention Logic* (John Wiley and Sons, Hoboken, NJ, 2005).
Feinberg, G. and Shapiro, R., 1980, *Life Beyond Earth: The Intelligent Earthlings Guide to Life in the Universe* (William Morrow and Company, New York, 1980).
Gelfand, I. M., Fomin, S. V., Silverman, R. A. (tr), 2000, *Calculus of Variations* (Dover Publications, Mineola, NY, 2000).
Giaquinta, M., Modica, G., Souchek, J., 1998, *Cartesian Coordinates in the Calculus of Variations* Volumes I and II (Springer-Verlag, New York, 1998).
Giaquinta, M., Hildebrandt, S., 1996, *Calculus of Variations* Volumes I and II (Springer-Verlag, New York, 1996).
Gradshteyn, I. S. and Ryzhik, I. M., 1965, *Table of Integrals, Series, and Products* (Academic Press, New York, 1965).
Heitler, W., 1954, *The Quantum Theory of Radiation* (Claendon Press, Oxford, UK, 1954).
Huang, Kerson, 1992, *Quarks, Leptons & Gauge Fields 2nd Edition* (World Scientific Publishing Company, Singapore, 1992).
Jost, J., Li-Jost, X., 1998, *Calculus of Variations* (Cambridge University Press, New York, 1998).
Kaku, Michio, 1993, *Quantum Field Theory*, (Oxford University Press, New York, 1993).
Kirk, G. S. and Raven, J. E., 1962, *The Presocratic Philosophers* (Cambridge University Press, New York, 1962).
Landau, L. D. and Lifshitz, E. M., 1987, *Fluid Mechanics 2nd Edition*, (Pergamon Press, Elmsford, NY, 1987).
Rescher, N., 1967, *The Philosophy of Leibniz* (Prentice-Hall, Englewood Cliffs, NJ, 1967).
Riesz, Frigyes and Sz.-Nagy, Béla, 1990, *Functional Analysis* (Dover Publications, New York, 1990).
Sakurai, J. J., 1964, *Invariance Principles and Elementary Particles* (Princeton University Press, Princeton, NJ, 1964).
Weinberg, S., 1972, *Gravitation and Cosmology* (John Wiley and Sons, New York, 1972).
Weinberg, S., 1995, *The Quantum Theory of Fields Volume I* (Cambridge University Press, New York, 1995).

INDEX

About the Author

Stephen Blaha is a well-known Physicist and Man of Letters with interests in Science, Society and civilization, the Arts, and Technology. He had an Alfred P. Sloan Foundation scholarship in college. He received his Ph.D. in Physics from Rockefeller University. He has served on the faculties of several major universities. He was also a Member of the Technical Staff at Bell Laboratories, a manager at the Boston Globe Newspaper, a Director at Wang Laboratories, and President of Blaha Software Inc. and of Janus Associates Inc. (NH).

Among other achievements he was a co-discoverer of the "r potential" for heavy quark binding developing the first (and still the only demonstrable) non-Aeolian gauge theory with an "r" potential; first suggested the existence of topological structures in superfluid He-3; first proposed Yang-Mills theories would appear in condensed matter phenomena with non-scalar order parameters; first developed a grammar-based formalism for quantum computers and applied it to elementary particle theories; first developed a new form of quantum field theory without divergences (thus solving a major 60 year old problem that enabled a unified theory of the Standard Model and Quantum Gravity without divergences to be developed); first developed a formulation of complex General Relativity based on analytic continuation from real space-time; first developed a generalized non-homogeneous Robertson-Walker metric that enabled a quantum theory of the Big Bang to be developed without singularities at t = 0; first generalized Cauchy's theorem and Gauss' theorem to complex, curved multi-dimensional spaces; received Honorable Mention in the Gravity Research Foundation Essay Competition in 1978; first developed a physically acceptable theory of faster-than-light particles; first derived a composition of extremums method in the Calculus of Variations; first quantitatively suggested that inflationary periods in the history of the universe were not needed; first proved Gödel's Theorem implies Nature must be quantum; provided a new alternative to the Higgs Mechanism, and Higgs particles, to generate masses; first showed how to resolve logical paradoxes including Gödel's Undecidability Theorem by developing Operator Logic and Quantum Operator Logic; first developed a quantitative harmonic oscillator-like model of the life cycle, and interactions, of civilizations; first showed how equations describing superorganisms also apply to civilizations. A recent book shows his theory applies successfully to the past 14 years of history and to *new* archaeological data on Andean and Mayan civilizations as well as Early Anatolian and Egyptian civilizations.

He first developed an axiomatic derivation of the form of The Standard Model from geometry – space-time properties – The Unified SuperStandard Model. It unifies all the known forces of Nature. It also has a Dark Matter sector that includes a Dark ElectroWeak sector with Dark doublets and Dark gauge interactions. It uses quantum coordinates to remove infinities that crop up in most interacting quantum field theories and additionally to remove the infinities that appear in the Big Bang and generate inflationary growth of the universe. It shows gravity has a MOND-like form without sacrificing Newton's Laws. It relates the interactions of the MOND-like sector of gravity with the r-potential of Quark Confinement. The axioms of the theory lead to the question of their origin. I suggest in the preceding edition of this book it can be attributed to an entity with God-like properties. I explore these properties in "God Theory" and show they predict that the Cosmos exists forever although individual universes (or incarnations of our universe) "come and go." Several other important results emerge from God Theory such a functionally triune God. The Unified SuperStandard Theory has many other important parts described in the Current Edition of *The Unified SuperStandard Theory* and expanded in subsequent volumes.

Blaha has had a major impact on a succession of elementary particle theories: his Ph.D. thesis (1970), and papers, showed that quantum field theory calculations to all orders in ladder approximations could not give scaling deep inelastic electron-nucleon scattering. He later showed the eigenvalue equation for the fine structure constant α in Johnson-Baker-Willey QED had a zero at $\alpha = 1$ not 1/137 by solving the Schwinger-Dyson equations to all orders in an approximation that agreed with exact results to 4th order in α thus ending interest in this theory. In 1979 at Professor Ken Johnson's (MIT) suggestion he calculated the proton-neutron mass difference in the MIT bag model and found the result had the wrong sign reducing interest in the bag model. These results all appear in Physical Review papers. In the 2000's he repeatedly pointed out the shortcomings of SuperString theory and showed that The Standard Model's form could be derived from space-time geometry by an extension of Lorentz transformations to faster than light transformations. This deeper space-time basis greatly increases the possibility that it is part of THE fundamental theory. Recently, Blaha showed that the Weak interactions differed significantly from the Strong, electromagnetic and gravitation interactions in important respects while these interactions had similar features, and suggested that ElectroWeak theory, which is essentially a glued union of the Weak interactions and Electromagnetism, possibly modulo unknown Higgs particle features, be replaced by a unified theory of the other interactions combined with a stand-alone Weak interaction theory. Blaha also showed that, if Charmonium calculations are taken seriously, the Strong interaction coupling constant is only a factor of five larger than the electromagnetic coupling constant, and thus Strong interaction perturbation theory would make sense and yield physically meaningful results.

In graduate school (1965-71) he wrote substantial papers in elementary particles and group theory: The Inelastic E- P Structure Functions in a Gluon Model. Phys. Lett. B40:501-502,1972; Deep-Inelastic E-P Structure Functions In A Ladder Model With Spin 1/2 Nucleons, Phys.Rev. D3:510-523,1971; Continuum Contributions To The Pion Radius, Phys. Rev. 178:2167-2169,1969; Character Analysis of U(N) and SU(N), J. Math. Phys. 10, 2156 (1969); and The Calculation of the Irreducible Characters of the Symmetric Group in Terms of the Compound Characters, (Published as Blaha's Lemma in D. E. Knuth's book: *The Art of Computer Programming Vols. 1 – 4*).

In the early 1980's Blaha was also a pioneer in the development of UNIX for financial, scientific and Internet applications: benchmarked UNIX versions showing that block size was critical for UNIX performance, developing financial modeling software, starting database benchmarking comparison studies, developing Internet-like UNIX networking (1982) and developing a hybrid shell programming technique (1982) that was a precursor to the PERL programming language. He was also the manager of the AT&T ten-year future products development database. His work helped lead to commercial UNIX on computers such as Sun Micros, IBM AIX minis, and Apple computers.

In the 1980's he pioneered the development of PC Desktop Publishing on laser printers and was nominated for three "Awards for Technical Excellence" in 1987 by PC Magazine for PC software products that he designed and developed.

Recently he has developed a theory of Megaverses – actual universes of which our universe is one – with quantum particle-like properties based on the Wheeler-DeWitt equation of Quantum

Gravity. He has developed a theory of a baryonic force, which had been conjectured many years ago, and estimated the strength of the force based on discrepancies in measurements of the gravitational constant G. This force, operative in D-dimensional space, can be used to escape from our universe in "uniships" which are the equivalent of the faster-than-light starships proposed in the author's earlier books. Thus travel to other universes, as well as to other stars is possible.
Blaha also considered the complexified Wheeler-DeWitt equation and showed that its limitation to real-valued coordinates and metrics generated a Cosmological Constant in the Einstein equations.
The author has also recently written a series of books on the serious problems of the United States and their solution as well as a book on the decline of Mankind that will follow from current social and genetic trends in Mankind.

In the past twenty years Dr. Blaha has written over 80 books on a wide range of topics. Some recent major works are: *From Asynchronous Logic to The Standard Model to Superflight to the Stars*, *All the Universe!*, *SuperCivilizations: Civilizations as Superorganisms*, *America's Future: an Islamic Surge, ISIS, al Qaeda, World Epidemics, Ukraine, Russia-China Pact, US Leadership Crisis*, *The Rises and Falls of Man – Destiny – 3000 AD: New Support for a Superorganism MACRO-THEORY of CIVILIZATIONS From CURRENT WORLD TRENDS and NEW Peruvian, Pre-Mayan, Mayan, Anatolian, and Early Egyptian Data, with a Projection to 3000 AD*, and *Mankind in Decline: Genetic Disasters, Human-Animal Hybrids, Overpopulation, Pollution, Global Warming, Food and Water Shortages, Desertification, Poverty, Rising Violence, Genocide, Epidemics, Wars, Leadership Failure.*

He has taught approximately 4,000 students in undergraduate, graduate, and postgraduate corporate education courses primarily in major universities, and large companies and government agencies.

He developed a quantum theory, The Unified SuperStandard Theory (UST), which describes elementary particles in detail without the difficulties of conventional quantum field theory. He found that the internal symmetries of this theory could be exactly derived from an octonion theory called QUeST. He further found that another octonion theory (UTMOST) describes the Megaverse. It can hold QUeST universes such as our own universe. It has an internal symmetry structure which is a superset of the QUeST internal symmetries.

Recently he developed Octonion Cosmology. He replaced it with HyperCosmos theory, which has significantly better features. He developed a fractionalization process for dimensions, particles and symmetry groups. He also described transformation that reduced particles and dimensions to a far more compact form. He also developed a precursor theory ProtoCosmos that leads to the HyperCosmos.

The author showed that space-time and Internal Symmetries can be unified in any of the ten HyperCosmos spaces in their associated HyperUnification spaces. The combined set of HyperUnification spaces enable all HyperCosmos dimensions to be obtained by a General Relativistic transformation from one primordial dimension in the 42 space-time dimension unified HyperUnification space.

At present the author devel;oped the Cosmos Theory that incorporates ProtoCosmos Theory, HyperCosmos Theory, Limos Theory, Second Kind HyperCosmos Theory and HyperUnification Spaces. He has introduced PseudoFermion wave functions and theory, He has related Cosmos Theory to Regge trajectories of spaces, parton theory, Veneziano amplitudes,

Fibonacci numbers and Ramsey numbers. He has calculated an approximation to the difficult R(n,n) Ramsey numbers.

He has developed a Gambol Model that successfully accounts for e-p deep inelastic scattering, fundamental particle resonances, hadron scattering, and the inner structure of particles based on confinement through Casimir forces of ideal gambol gases. The Gambol Planckian Distribution was derived.

He has applied the Gambol Model to particles, universes, and the Cosmos of universes. He showed that the Cosmos may have a distribution of 23 universes corresponding to various Cosmos spaces.

Recently he showed that Cosmos Theory follows from the number of independent asymmetric tensors in a dimension r. He also showed the close parallel between the form of γ-matrices and Cosmos Theory dimension arrays. The closeness suggested that dimension arrays have the same importance as γ-matrices for fermions.

He demonstrated that the pressure of fermions within a space of dimension r balances the Casimir vacuum energy force for 18 dimensions. He showed that $2e\pi = 17.02$ marks the critical point where pressure balances Casimir force, which implies $r = 18$ is the highest dimension Physical Cosmos space. The dimension $2e\pi$ appears to set the approximate dimension for Cosmos spaces with dimension array size $2^{r+4} \cong (17.02/8)^{r+4}. \cong (e\pi/4)^{r+4}. \cong 2.13^{r+4}$.

Now he has found the sequences of Coupling Constant values and fermion masses in the Standard Model and UST. He unified the Cosmos spaces spectrum, the coupling constant spectrum and the fermion mass sequences in a fundamental scaling differential equation.

The author's Cosmos Theory is now well explored. This book develops Gravitation and a Theory of Time within Cosmos Theory.